ADAPTATIONS r PHYSIOL

— MELODY CLARK.

AF851259

Molluscan Shellfish Farming

Molluscan Shellfish Farming

by

B. E. Spencer

Fishing News Books

Blackwell
Science

Editorial Offices:
Osney Mead, Oxford OX2 0EL, UK
Tel: +44 (0)1865 206206
Blackwell Science, Inc., 350 Main Street, Malden, MA 02148-5018, USA
Tel: +1 781 388 8250
Iowa State Press, a Blackwell Publishing Company, 2121 State Avenue, Ames, Iowa 50014-8300, USA
Tel: +1 515 292 0140
Blackwell Publishing Asia Pty Ltd, 550 Swanston Street, Carlton South, Victoria 3053, Australia
Tel: +61 (0)3 9347 0300
Blackwell Wissenschafts Verlag, Kurfürstendamm 57, 10707 Berlin, Germany
Tel: +49 (0)30 32 79 060

First published 2002

Library of Congress
Cataloging-in-Publication Data
is available

ISBN 0-85238-291-X

A catalogue record for this title is available from the British Library

Set in 11 on 13pt Bembo
by SNP Best-set Typesetter Ltd., Hong Kong
Printed and bound in Great Britain by
MPG Books Ltd, Bodmin, Cornwall

For further information on
Blackwell Science, visit our website:
www.blackwell-science.com

Contents

Frank Buckland and the Buckland Foundation

The Buckland Foundation was endowed by Frank Buckland, a sometime surgeon, natural historian, fisheries scientist and government inspector of salmon fisheries, but above all an inveterate communicator of the period 1860–80. People flocked to hear him talk and he never let slip any opportunity to make people aware of the importance of the fisheries and their problems in particular and of natural history in general.

Buckland trained as a surgeon and took a commission in the Life Guards in 1854, but from the outset began to write popular articles on natural history. These were published in 1857 to great acclaim as a book entitled *Curiosities of Natural History*, which grew to four volumes and appeared in 15 editions. His success increased demands upon him as a writer and lecturer and he resigned his commission in 1863.

He had become interested in fish culture, then regarded simply as the rearing of fish from the egg, and an exhibit he had in the office window of *The Field* magazine on The Strand in London proved to be so popular that it stopped the traffic on occasion. He gave a successful lecture on the subject in 1863 at the Royal Institution, subsequently published as *Fish Hatching*, and was struck by the intense interest that it aroused. One consequence of it was that he set up a small aquarium at the South Kensington Museum, the forerunner of the Science Museum, and by 1865 had collected there a range of exhibits which were to form the nucleus of his eventual 'Museum of Economic Fish Culture.' This aimed to inform the public about the fish and fisheries of the British Isles, and for the rest of his life he laboured, in part, to develop this display. At about the same time (1867) he was appointed to the office of Board of Trade Salmon Inspector and he also set up the weekly magazine *Land and Water*.

Britain's growing population in the nineteenth century created many problems of food supply. The sea fisheries offered a cheap source of abundant first class protein, and as a result of the latent demand and the growth of railways the North Sea fisheries grew spectacularly. Little was known about sea fish: no statistics of fish landings were available, at least in England, and there was little formal research. Buckland sat on four Commissions that looked at Fish and Fishing between 1875 and his death in 1880. In the same period a number of public fisheries exhibitions were held abroad and he tirelessly pressed for something similar to be staged in the United Kingdom. At all times he was concerned to explain, to teach and, most particularly, to make the general public aware of the importance of the fisheries and the need to protect and develop this great national asset. 'Buckland's voice,' says Geoffrey Burgess, his most recent biographer, 'remains the only one consistently calling for research into fishery problems, publicising the activities of the industry, drawing attention to the national importance of fish in the diet, and acting as a focus for those in the industry and elsewhere who were interested and concerned about its proper commercial development.'

Something of the flavour of his views is given by the following quotations from his reports and articles.

> 'A greater cry should more properly be established against those which deter or kill the fish by noxious materials which they pour into public waters for their private use and Benefit . . . The above mentioned individuals and companies reap no inconsiderable profits from their individual operations, but while endeavouring to increase their own profits, they treat with indifference the welfare of the public, and an important source of food not only to themselves, but to the public in general.'

> 'What objection can be reasonably argued against the employment of revenue cruisers for the accommodation of naturalists, appointed by government . . . in order that they make a thoroughly practical examination of the dark and mysterious habits of food fishes. The trawl and tow net, we firmly believe, if judiciously and persistently employed over an extended area of the sea, by men able to identify what the nets drag up and entangle, would do more to bring to light what is now hidden and unknown than all the evidence collected by the Sea Fisheries Commission. It is a government question, and not one of private of individual research. We feel confident that the time is not far distant when properly–equipped naturalists will be sent by government to investigate the habits of deep sea fish.'

> 'We want also samples of the surface water itself under peculiar conditions, for instance, what is the meaning of the wonderful white appearance of the sea which took place last autumn in nearly all the waters of the northern coast of England? What is the meaning of the occasional red appearance of the sea for many square miles? Again, how are we to devise a mesh of net that shall let go the small soles and undersized fry of other sea fish, and keep marketable fish only?'

Frank Buckland died in December 1880. A few days before his death he signed his will. His wife was to have a life interest in his estate but on her death £5000 would be used to establish a trust fund to support 'A professorship of Economic Fish Culture, to be called The Buckland Professorship'. The money became available in 1925 and in 1930 the first Buckland Lectures were given. It is clear that Frank Buckland intended the term 'Fish Culture' to be widely interpreted and to cover much more than fish hatching and the rearing of fry.

The invitation to deliver the Buckland Lectures is made by the Board of Trustees: one representing the Department of Environment, Food and Rural Affairs and two the Department of Education and Employment, the direct descendent of the Ministry of Education of 1925. Down the years they have sought to ensure that the lectures are not only of interest and relevance to as broad a section of society as possible but also of value to those who depend for their livelihood on some aspect of fish and fishing. The Trustees hope that by continuing to keep alive the memory of a man who dedicated his life to the improvement of the commercial fisheries of the British Isles and help, in their turn, to improve people's understanding of current problems in the commercial fisheries and their environment.

Further Reading

Burgess, G. H. O. (1967) *The Curious World of Frank Buckland.* John Baker, London.

List of Buckland Lectures and Professors

1930	*Frank Buckland's Life and Work*	W. Garstang
1931	*Salmon Hatching and Salmon Migrations*	W. L. Calderwood
1932	*The Natural History of the Herring in Scottish Waters*	H. Wood
1933	*The Natural History of the Herring in the Southern North Sea*	W. C. Hodgson
1934	*The Hake and the Hake Fishery*	C. F. Hickling
1935	*Oyster Biology and Oyster Culture*	J. H. Orton
1936	*The Nation's Fish Supply*	E. Ford
1937	*Fish Passes*	T. E. Pryce Tarrant
1938	*Hydrography in Relation to Fisheries*	J. B. Tait
1939	*Rational Fishing of the Cod in the North Sea*	M. Graham
1947	*The Stock of Salmon, its Migrations, Preservation and Improvement*	W. J. Menzies
1948	*Sea Fisheries*	G. T. Atkinson
1949	*The Plaice*	R. S. Wimpenny
1950	*River Pollution*	H. D. Turing
1951	*Fishery Hydrography*	J. R. Lumby
1952	*River Purification*	F. T. K. Pentelow
1953	*Irish Salmon and Salmon Fisheries*	A. E. J. Went
1954	*Inshore Fisheries*	H. A. Cole
1955	No appointment	

1956	*The Haddock*	B. B. Parrish
1957	*Plankton* (published as *The Fertile Sea*)	A. P. Orr
1958	*Lemon Sole*	A. R. Bennett
1959	*Fish Capture*	R. Balls
1960	*Historical Background of International Organisations for Regulating Fisheries, their achievements so far, and prospects for the future*	R. J. H. Beverton
1961	*The Stocks of Whales*	N. A. Mackintosh
1962	No appointment	
1963	*British Freshwater Fishes*	M. E. Varley
1964	*Developments in the Handling and Processing of Fish*	G. H. O. Burgess
1965	*The Lobster – its Biology and Fishery*	H. J. Thomas & A. C. Simpson
1966	*Sonar in Fisheries – a Forward Look*	D. G. Tucker
1967	*The Artificial Cultivation of Shellfish*	P. R. Walne
1968	No appointment	
1969	*Ocean Currents and their Influence on Fisheries*	A. J. Lee
1970	*Fish, Nets and Men – an Underwater Approach to Fisheries Research*	C. C. Hemmings
1971	*Behaviour and the Fisheries*	F. R. Harden Jones
1972	No appointment	
1973	No appointment	
1974	*Exploitation of the Salmon Stocks*	K. A. Pyefinch
1975	No appointment	
1976	No appointment	
1977	*The Edible Crab and its Fishery*	E. Edwards
1978	No appointment	
1979	*Maximum Use of British Aquatic Food Resources*	J. J. Connell & R. Hardy
1980	*Scallop and Queen Fisheries in the British Isles*	J. Mason
1981	*Marine Pollution and its Effect on Fisheries*	A. Preston & P. C. Wood
1982	*Engineering, Economics and Fisheries Management*	G. Eddie
1983	*A Story of the Herring*	G. Buchan
1984	*The Mackerel – its Biology, Assessment and the Management of a Fishery*	S. J. Lockwood
1985	*Aquaculture*	R. J. Roberts
1986	No appointment	
1987	*The Norway Lobster*	C. J. Chapman
1988	*North Sea Cod*	C. T. Macer
1989	*Atlantic Salmon*	W. M. Shearer
1990	*Pollution and Freshwater fisheries*	R. Lloyd
1991	*The Common Fisheries Policy: Past, Present and Future*	M. J. Holden
1992	*Industrial Fisheries, Fish Stocks and Seabirds*	R. S. Bailey
1993	*Marine Protected Areas and Fisheries*	S. Gubbay
1994	*Deep-Sea Fisheries: a new resource?*	J. D. M. Gordon
1995	*Bivalve Cultivation in the UK: Structuring Influences*	B. E. Spencer
1996	*Protecting the Seas: using Science for a Better Environment*	J. S. Gray
1997	*Global Change in the Coastal Zone – Implications for Fisheries*	P. Holligan

1998	*The Deep Ocean: Use and Misuse*	M. Angel
1999	*The European Eel*	C. Moriarty
2000	*Integrated fisheries management – a challenge for the Common Fisheries Policy*	D. Symes
2001	*Plankton, fish and climate change*	P.C. Reid

Most of the lectures have been published in book form; many are still in print. For more information as to their availability and also for a list of Buckland Occasional Papers please contact: John Ramster, Clerk to the Buckland Foundation, 3 Woodside Avenue, Bridge of Weir, PA11 3PQ. Phone/fax 01505 615402, email jramster@lineone.net

Acknowledgements

Brian Spencer and Blackwell Publishing would like to thank the following people and publishers for permission to use their material in the figures, text and tables cited below.

Chapter 1

Figure 1.10 photograph: D. LL. Hugh-Jones.

Figure 1.12 from: Dare, P. J., (1993) *Biological considerations in the exploitation of mussels*. The Dr Walne Memorial Lecture, in the Proceedings of the 24th Annual Conference of the Shellfish Association of Great Britain, Fishmongers Hall, London.

Chapter 2

Figure 2.1 adapted from Fig. 2.2 in: Morton, B. (1992) The evolution and success of the heteromyarian form in the Mytiloida. In: *The mussel,* Mytilus: *Ecology, Physiology, Genetics and Culture*. (ed E. Gosling), pp. 21–52. Developments in Aquaculture and Fisheries Science, 25, Elsevier, Amsterdam.

Figure 2.2 adapted from Fig. 82 in: Yonge, C. M. & Thompson, T. E. (1976) *Living Marine Molluscs*. Collins, London.

Figure 2.3 & Figure 2.4 adapted from: Kennedy, V. S., Newell, R. I. E. & Eble, A. F. (eds) (1996) *The Eastern Oyster*, Crassostrea virginica. Maryland Sea Grant College, USA.

Figure 2.5 adapted from Fig. 8 in: Ward, J. E., Newell, R. I. E., Thompson, R. J. & McDonald, B. A. (1994) *In vivo* studies of suspension-feeding processes in the eastern oyster, *Crassostrea virginica* (Gmelin). *Biological Bulletin*, **186**, 221–40.

Figure 2.6 adapted from Fig. 1 in: Newell, R. I. E. & Langdon, C. J. (1996) Mechanisms and physiology of larval and adult feeding. In: (eds V. Kennedy, S., R. I. E. Newell & A. F. Eble), 1996 *The Eastern Oyster*, Crassostrea virginica. Maryland Sea Grant College, USA.

Figure 2.7 redrawn from: Vahl, O. (1973) Efficiency of particle retention in *Chlamys islandica*. *Astarte*, **6**, 21–25.

Chapter 3

All figures have been redrawn from those originally published in: Utting, S. D. & Spencer, B. E. (1991) *The hatchery culture of bivalve mollusc larvae and juveniles*. Laboratory Leaflet, MAFF Directorate of Fisheries Research, Lowestoft, 68. Copyright permission granted by CEFAS (MAFF publication).

Chapter 4

All figures have been redrawn from those originally published in: Spencer, B. E. (1990) *Cultivation of Pacific Oysters*. Laboratory Leaflet, MAFF Directorate of Fisheries Research, Lowestoft, 63. Copyright permission granted by CEFAS (MAFF publication).

Chapter 5

All figures have been redrawn from those originally published in: Spencer, B. E., Edwards, D. B. & Millican, P. F. (1991) *Cultivation of Manila Clams*. Laboratory Leaflet, MAFF Directorate of Fisheries Research, Lowestoft, 65. Copyright permission granted by CEFAS (MAFF publication).

Chapter 6

Figures 6.5, 6.6, 6.7 adapted from: Cai, Y. & Li X. (1990) Oyster culture in the People's Republic of China. *World Aquaculture* **21**, 467–72.

Figure 6.8 adapted from: Berthomé, J-P., Prou, J. Razet, D. & Garnier, J. (1984) First approach of provisional estimation of the potential production of the cupped oyster, *Crassostrea gigas*. *Haliotis*, **14**, 39–48.

Chapter 7

Figure 7.3, 7.4 adapted from: Hickman, R. W. (1992) Mussel Cultivation. In: *The mussel,* Mytilus: *Ecology, Physiology, Genetics and Culture* (ed E. Gosling) Developments is Aquaculture and Fisheries Science, 25, pp. 465–510 Elsevier, Amsterdam

Figure 7.5 photograph from: Xplora Products Ltd., Unit 2, Greenelms Trading Estate, Greys Rd., Uddingston, Glasgow G71 6ET.

Figure 7.8 redrawn from: (Davies, G., Dare, P. J. & Edwards, D. B. (1980) *Fenced enclosures for the protection of seed mussels* (Mytilus edulis *L.*) *from predation by shore-crabs* (Carcinus maenas (*L*). MAFF, Directorate of Fisheries Research, Lowestoft, England, Fisheries Research Technical Report, 56. Copyright permission granted by CEFAS (MAFF publication).

Figure 7.9 photograph from: Sea Fish Industry Authority, Marine Farming Unit, Ardtoe, Acharacle, Argyll PH36 4LD.

Table 7.2 & Table 7.4 data derived from: Hickman, R. W. (1992) Mussel Cultivation. In: *The mussel* Mytilus: *Ecology, Physiology, Genetics and Culture.* (ed E. Gosling), Developments in Aquaculture and Fisheries Science, 25, pp. 465–510 Elsevier, Amsterdam.

Table 7.5 data derived from: Dare, P. J. (1980) *Mussel Cultivation in England and Wales.* Laboratory Leaflet, MAFF, Directorate of Fisheries Research, Lowestoft, England. 50. Copyright permission granted by CEFAS (MAFF publication). Also from: Stirling, H. P. & Okumus, I. (1995). Growth and production of mussels (*Mytilus edulis* L.) suspended at salmon cages and shellfish farms in two Scottish sea lochs. *Aquaculture*, **134**: 193–210.

Chapter 8

Figures 8.1, 8.2, 8.4, and 8.5 adapted from: Ventilla, R.F. (1982) The Scallop Industry in Japan. In: *Advances in Marine Biology*, 20 (eds J. H. S. Blaxter, F. S. Russell & C. M. Yonge), pp. 309–82. Academic Press, London and New York

Figures 8.3 and 8.6 adapted from: Susumu Ito (1991) Patinopecten (Mizuhopecten) yessoensis *(Jay) in Japan.* In: *Estuarine and Marine Bivalve Mollusk Culture*, (ed W. Menzel), pp. 211–225 CRC Press Inc., Boston.

Figures 8.7 and 8.8 photographs from: SFIA, Marine Farming Unit, Ardtoe, Acharacle, Argyll PH36 4LD.

Figures 8.9, 8.10, 8.11 and 8.12 redrawn from: data in (White Fish Authority (WFA), (1981). *An outline of the application of Japanese scallop culture techniques in Scotland and a review of recent WFA scallop culture research.* WFA Field Report No. 937, White Fish Authority, Edinburgh. (SFIA, Marine Farming Unit, Ardtoe, Acharacle, Argyll PH36 4LD.)

Table 8.2 data from: Dare, P. J. (1996) *Scallop cultivation in the UK: a desktop study of R & D progress with* Pecten maximus *during 1975–1994.* Report to the Ministry of Agriculture, Fisheries and Food, Chief Scientist's Group, London. Contract No. Ref. CSA 2967.

Chapter 9

Figures 9.2 & 9.3 adapted from: Tong, L. J., Moss, G. A., Redfearn, P. & Illingworth, J. (1992) *A manual of techniques for culturing paua,* Haliotis iris, *through the early juvenile stage*. New Zealand Fisheries Technical Report, 31. Fisheries Research Division, Wellington, New Zealand.

Figure 9.4 adapted from: Tong, L J. (1982) The potential of aquaculture of paua in New Zealand. In: *Proceedings of the Paua Fishery Workshop*. (Comps Akroyd, J. M., T. E., Murray J. L. Taylor) Fisheries Research Division Occasional Publication 41, pp. 36–40. Fisheries Research Division, Wellington, New Zealand.

Figure 9.5 redrawn from: Chen, H-C. (1989) Farming the small abalone, *Haliotis diversicolor supertexa*, in Taiwan. In: *Handbook of Culture of Abalone and Other Marine Gastropods*. (ed K. O. Hahn) pp. 265–83. CRC Press, Florida.

Figure 9.6 redrawn from: Sung, K. Y. (1989) Abalone farming in Korea. In: *Handbook of Culture of Abalone and Other Marine Gastropods*. (ed K. O. Hahn) pp. 255–63. CRC Press, Florida, USA.

Figure 9.7 from data in: La Touche, R., Moyan, K. & Twomey, W. (1993) *Abalone on-growing manual*. Aquaculture Explained, 14, Bord Iascaigh Mhara, Dublin.

Table 9.1 adapted from: Jarayabhand, P. & Paphavasit N. (1996) A review of the culture of tropical abalone with special reference to Thailand. *Aquaculture*, **140**, 159–68. Also: Fallu, N. (1991) *Abalone Farming*, Fishing News Books, Blackwell Science Ltd Oxford; Hahn, K. O. (1989) (ed) *Handbook of Culture of Abalone and Other Marine Gastropods*. CRC Press, Florida, USA.

Table 9.3 and Table 9.6 from: La Touche, R., Moyan, K. & Twomey, W. (1993) *Abalone on-growing manual*. Aquaculture Explained, 14, Bord Iascaigh Mhara, Dublin.

Table 9.4 from: Flemming, A. E., Van Barneveld, R. J. & Hone, P. W. (1996) The development of artificial diets for abalone: a review and future directions. *Aquaculture*, **140**, 5–53.

Table 9.5 from: Aviles, J. G. G. & Shepherd, S. A. (1996) Growth and survival of the blue abalone *Haliotis fulgens* in barrels at Cedros Island, Baja California, with a review of abalone barrel culture. *Aquaculture*, **140**, 169–76.

Chapter 10

Text from: Spencer, B. E. (1991) Predators and methods of control in molluscan shellfish cultivation in north European waters, pp. 309–37. In:

Aquaculture and the Envirionment. (eds, N. De Pauw & J. Joyce). European Aquaculture Society Special Publication No. 16, Gent, Belgium. Copyright permission for text granted by CEFAS (EAS publication).

Figure 10.1 (a) from: Dare, P. J., Davies, G. & Edwards, D. B. (1983) *Predation on juvenile Pacific oysters* (Crassostrea gigas *Thunberg*) *and mussels* (Mytilus edulis *L.*) *by shore crabs*, (Carcinus maenas (*L.*)). Fisheries Research Technical Report, MAFF, Directorate Fisheries Research, Lowestoft **73**: 1–15; (b) Walne, P. R. & Dean, G. J. (1972) Experiments on predation by the shore crab, *Carcinus maenas* L., on *Mytilus* and *Mercenaria. Journal du Conseil international pour l'Exploration de la* Mer **34** (2), 190–99. Copyright permission granted by CEFAS (MAFF publication). (c) Elner, R.W. (1980) The influence of temperature, sex and chela size on the foraging strategy of the shore crab, (*Carcinus maenas* (L.) *Marine Behaviour and Physiology* **7**: 15–24.

Figure 10.2 (a) from: Dare, P. J., Davies, G. & Edwards, D. B. (1983) *Predation on juvenile Pacific oysters* (Crassostrea gigas *Thunberg*) *and mussels* (Mytilus edulis *L.*) *by shore crabs*, (Carcinus maenas (*L.*)). Fisheries Research Technical Report, MAFF, Directorate Fisheries Research, Lowestoft **73**, 1–15. Copyright permission granted by CEFAS (MAFF publication); (b) Walne, P. R., & Dean, G. J. (1972) Experiments on predation by the shore crab, *Carcinus maenas* L., on *Mytilus* and *Mercenaria. Journal du Conseil international pour l'Exploration de la Mer* **34** (2), 190–99; (c) drawn from Table 1 in: Parache, A. (1980) *Les relations 'proie-predateur' entre le crabe verte* Carcinus maenas *et la palourde* Ruditapes philippinarum. Bulletin Mensuel, Office National de la Chasse, France, pp. 299–309; (d) Sanchez-Salazar, M. E., Griffiths,C. L. and Seed, R. (1987), The effect of size and temperature on the predation of cockles, *Cerastoderma edule* (L.) by the shore crab, *Carcinus maenas* (L.). *Journal of Experimental Marine Biology and Ecology*, **111**, 181–93; (e) Lake, N. C. H., Jones M. B. & Paul, J. D. (1987) Crab predation on scallop (*Pecten maximus*) and its implication for scallop cultivation. *Journal of the Marine Biological Association of the United Kingdom*, **67**, 55–64.

Figure 10.3 adapted from: IFREMER (1988) *La Palourde – dossier d'elevage.* IFREMER Documents and Publications Unit, Plouzane, France.

Figure 10.4 from: Davies, G., Dare, P. J. & Edwards, D. B. (1980) *Fenced enclosures for the protection of seed mussels* (Mytilus edulis (*L.*) *from predation by shore crabs* (Carcinus maenas (*L.*)). Fisheries Research Technical Report, MAFF Directorate of Fisheries Research, Lowestoft, 56 Copyright permission granted by CEFAS MAFF publication).

Figure 10.5 from: Korringa, P. (1976) (a) *Farming Marine Organisms Low in the Food Chain.* Vol. 1. *A Multi-disciplinary Approach to Edible Seaweed, Mussel and Clam Production*. Elsevier, Amsterdam. (b) *Farming the cupped oysters of the genus* Crassostrea vol. 2. A Multidisciplinary Treatise. Elsevier, Amsterdam.

Chapter 11

Text, tables and figures from: Spencer, B. E. (1990) *Cultivation of Pacific oysters*. Laboratory Leaflet, MAFF Directorate of Fisheries Research, Lowestoft, 63. Copyright permission granted by CEFAS (MAFF publication).

Chapter 12

Figure 12.3 from: Ayres P. A. (1978) *Shellfish purification in installations using ultraviolet light*. Laboratory Leaflet, MAFF Directoralie of Fisheries Research, Lowestoft, 43. Copyright permission granted by CEFAS (MAFF publication).

Table 12.1 from: Boulter, M. & Denton J. W. (1995) *Development of small-scale vertical stack oyster purification plant with repeated re-use of artificial sea water*. Seafish Report No. 459, Sea Fish Industry Authority, Hull, England.

Tables 12.3 and 12.4 from: Anon. 1995 *Seafish standard design purification systems: operating manual for the vertical stack system*. Technical Information Service, Sea Fish Industry Authority, Hull, England, Report, Reference No. 1995/32/FT-V1.

Table 12.6 from: Casagrande, F. (1988) Effective purification in aquaculture. *Marine Farming* also: Morel, M. & Salamon, W. (1992) Project of a depuration plant in northern France. In: *Purification des Coquillages*. Deuxième Conference International, April 1992, IFREMER, Rennes, France, pp. 323–36 and: Rodrick, G. E. & Schneider, K. R. (1992) Depuration and relaying of molluscan shellfish. pp. 331–63. In: *Environmental Indicators and Shellfish Safety*. (eds C. R. Hackney & M. D. Pierson) Chapman and Hall, New York.

Introduction

Bivalve cultivation as a Buckland lecture topic crops up every quarter of century or so, reflecting the need to update information and to measure progress of development in the intervening period. Since the previous series of lectures on bivalve cultivation given by Peter Walne in 1968, there has been less of a revolution in cultivation techniques but rather a period of consolidation of methodology. The 1960s saw great strides being made with the development of reliable hatchery methods. In the UK, Walne led the field with his work on understanding the feeding requirements of flat oyster larvae and spat and the application of this knowledge to high-density culture of oysters. Since that time, the basic hatchery techniques developed at Conwy, north Wales, and adopted worldwide have enabled countries to establish significant bivalve cultivation industries in the absence of reliable collection of wild spat, or to diversify with non-native or fringe species. Although hatcheries contribute significantly to world bivalve production, their contribution is overshadowed by production from fisheries based on the collection and on-growing of wild seed. Methods developed in Japan for the collection and growing of seed mussels, oysters and scallops, especially by suspended cultivation, have inspired many other countries to follow suit and to develop and refine the methodology to work in their own environments.

This book on molluscan shellfish cultivation is largely about bivalves but contains one chapter on abalone (a gastropod) cultivation. It is unlike those on the subject previously sponsored by the Buckland Trustees in not being based solely on a series of lectures about the work undertaken by the authors. In this book only Chapter 1 constitutes the Buckland lecture, and this is based on a personal assessment of the factors that have contributed to moulding the UK bivalve industry into its present-day structure. Some of the other chapters, describing the cultivation of bivalve larvae, on-growing of hatchery-reared oyster and clam spat, and site selection (Chapters 3, 4, 5 and 11) were co-authored and based on the authors' experiences and publications whilst working at the Fisheries Laboratory, Conwy, during the period 1960–1996. They were largely advisory (they were written in a style that reflects this) and developed as an aid to the UK shellfish industry to promote information to a readership that with all probability had little scientific/technical knowledge to draw upon.

I am deeply indebted to my co-authors, Sue Utting (Chapter 3), Brian Edwards and Peter Millican (Chapter 5) and Ian Laing (Chapter 11) for their major inputs in the original publications. One other chapter, on predators and their control (Chapter 10), is based on a lecture given by the author to the European Mariculture Society in 1992 and published in their journal. The remaining chapters were written to give the book wider appeal by encompassing species or topics that the molluscan shellfish cultivator is likely to meet anywhere in the world. I have relied heavily on published information to write these chapters, since most of them are on subjects outside of my personal experience. The source of information has sometimes been acknowledged in the text but mostly it has been inserted at the end of the chapter as a reference list. In some cases, particularly Chapters 2, 6–9, and 12, I have re-drawn, adapted or used drawings from published work and am pleased to acknowledge these where appropriate. Many of the chapters contain information of global catch and aquaculture statistics, published by the Food and Agriculture Organisation of the United Nations, or obtained from their web site (*http://www.fao.org/fi*) on the species under discussion. These data provide a measure of the performance of the main producer countries now and in the recent past.

I am deeply indebted to the Buckland Trustees, and in particular to Dr Stephen Lockwood, for his gentle persuasion in encouraging me to undertake this project and his continual encouragement during the writing phase. I am also indebted to my daughter, Julia Peters, for using her skills in computer graphics to re-draw many of the figures used in Chapters 2 and 3.

1 Bivalve cultivation in the UK: structuring influences

Introduction

This chapter, based on the Buckland Lecture of 1995, describes the status of bivalve culture in the world, in Europe and in the UK and the influences that have moulded the UK industry into its present state. The style of the lecture differed from the last one on bivalve cultivation, given in 1968 by Peter Walne, a respected pioneer of modern-day hatchery technology and inspiration to bivalve cultivation in the UK in the 1960s and 1970s, which concentrated on the science relating to the developing hatchery technology.

Production

World production

To set the scene, some aspects of world fisheries production are shown in Figs 1.1 and 1.2. Aquaculture production of fish, crustaceans, molluscs etc. was just below half that of fisheries capture production in 1999 and Asia was significantly the largest aquaculture producing continent (FAO, 2001). World mollusc production reached 15.9 million tonnes(t) in 1997 (Fig. 1.3), an increase of 21% from 1992, and 82% from 1987 (FAO, 1999). About half (54%) of this production, especially mussels, oysters and clams, originated directly from cultivated or managed fisheries, but cephalopods, representing 38% of the total production do not yet feature significantly in cultivated mollusc production. Whereas mussel production has remained relatively unchanged since 1992 (7% increase), clam and oyster production increased by 82% and 98%, respectively, and scallops, showing the largest increase, by 131%. In the league of top producers, China is at or near the top for all groups of cultivated species (Table 1.1), producing 87% of the 5.3 million tonnes increased production between 1992 and 1997. Other non-European countries such as Korea, Japan and the USA, featured prominently in world production of oysters, scallops and clams.

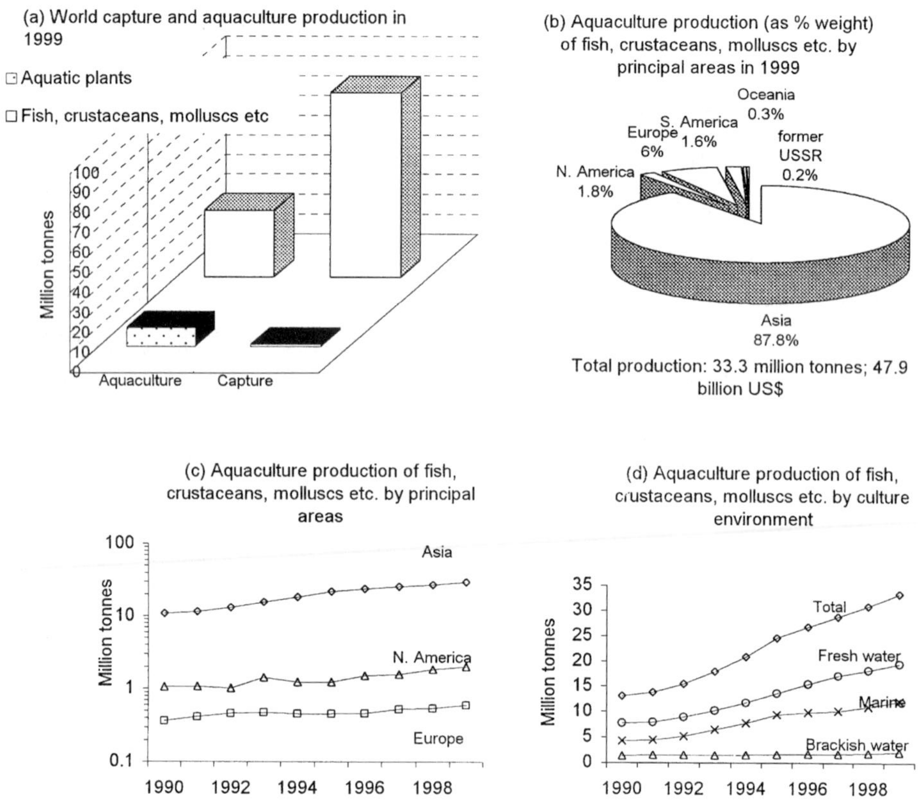

Fig. 1.1 Various world fisheries production values (FAO, 2001).

European production

Although China produced almost 400 000 t of mussels in 1997, some European countries also make significant contributions to world production. Spain and Italy each produced over 100 000 t of mussels, and the Netherlands, Denmark and France between 60 000 and 100 000 t each (Table 1.2). As far as oyster production is concerned, France was the major producer in Europe, with almost 150 000 t of Pacific oysters (*Crassostrea gigas*), a considerably greater production than its nearest rivals, Spain, Turkey and the Netherlands, each producing 1000–4000 t.

UK production

Bivalve production of cultivated species in the UK (mussels and oysters) comprises a relatively small proportion of that in Europe. Mussels (19 000 t)

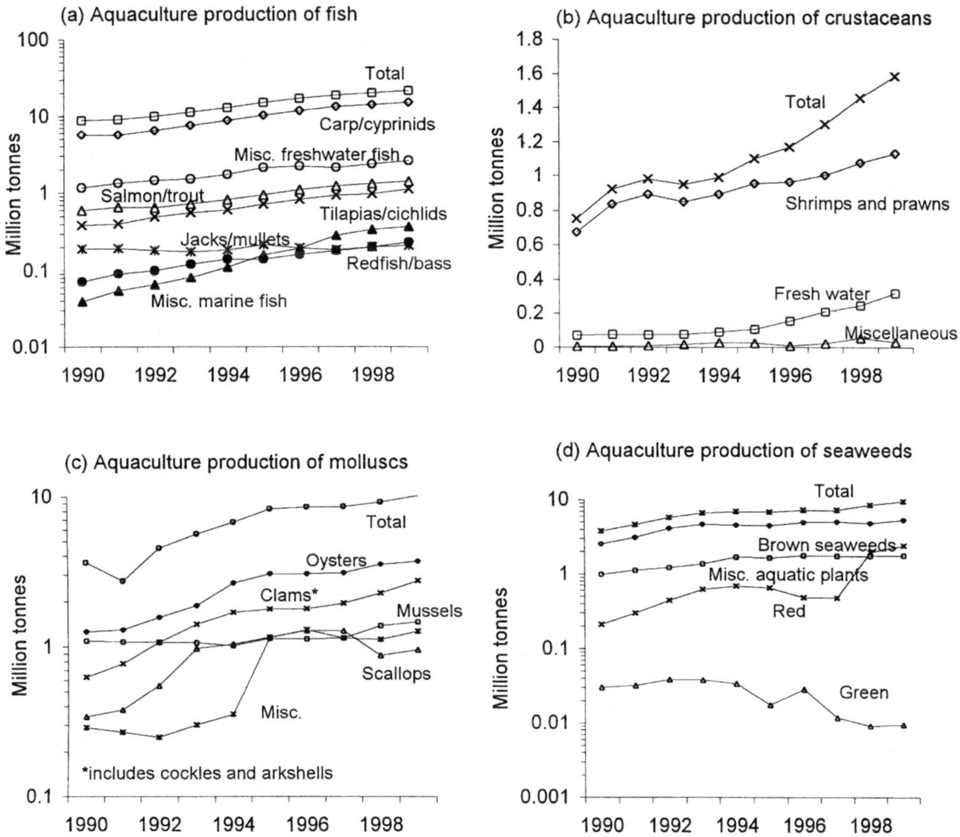

Fig. 1.2 Various world fisheries production values (FAO, 2001).

and oysters [1600 t of flat oysters (*Ostrea edulis*) and Pacific oysters] (Table 1.3) were about 3% and 1%, respectively, of European production in 1997.

In the UK mollusc landings totalled 80 900 t in 1997, worth £27.3 million (MAFF Sea Fisheries Statistical Tables; Table 1.3). The managed cockle fisheries produced 24% of the landings by weight and 13% by value. Scallops were the most valuable mollusc species with 18 500 t landed, worth £27.6 million. Although scallop cultivation is of prime interest in Scotland most of the landings are from natural stocks. The cultivated species, mussels and oysters, form *c.* 25% of the landings by weight and 18% by value.

The distribution of bivalve fisheries shows a major concentration on the east, south and south-west coasts of England and Wales and on the west coast of Scotland (Fig. 1.4). The development is influenced by the location of sheltered estuaries, bays and lochs. In England and Wales, the large natural fisheries are located in the Wash (cockles, mussels), the Thames estuary

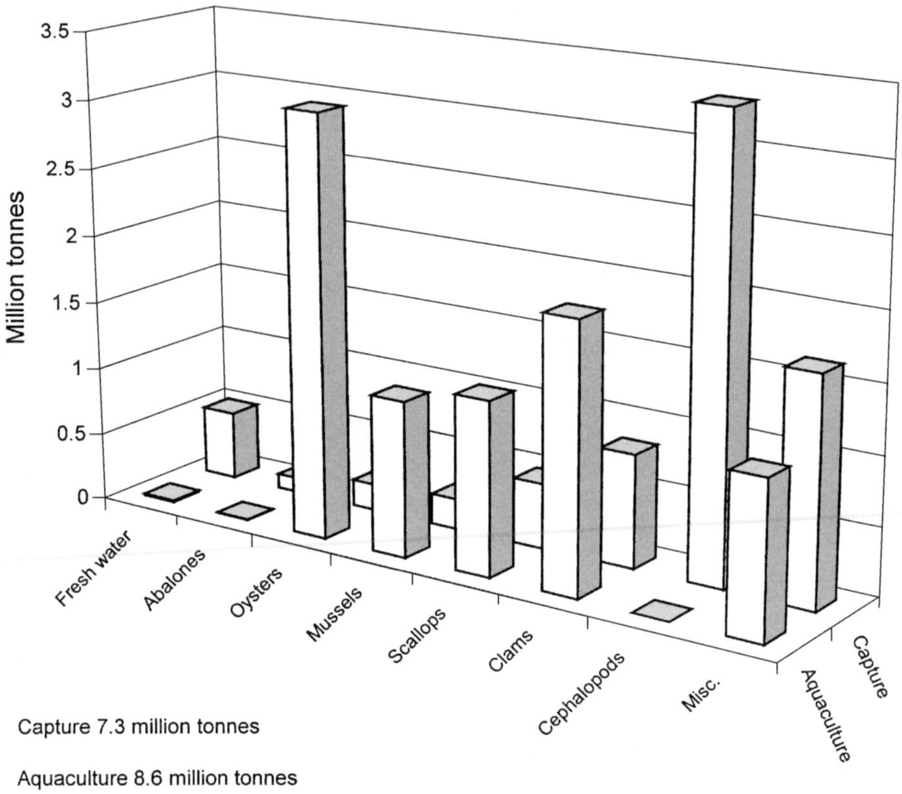

Fig. 1.3 World mollusc production from aquaculture and capture in 1997 (FAO, 1999).

(cockles), the Solent (oysters), the Fal (oysters), Burry Inlet (cockles), the Dee (cockles) and Morecambe Bay (cockles, mussels). Most of these are regulated fisheries, managed largely by Sea Fisheries Committees to ensure their sustainability. There are also a number of several fisheries for specified species of bivalve which give rights of ownership to individuals or groups of persons including Sea Fisheries Committees, Harbour Boards etc., who are required to manage and increase the production of the stocks. Many cultivators however have established sites leased from the Crown Estate Commission and/or from private owners of the foreshore.

Production from these farms, which numbered 110 in England and Wales and 170 in Scotland in 1997, show a general increase in recent years (Fig. 1.5). However, in recent decades forces that have been instrumental in moulding the present-day structure of the industry have influenced production both positively and negatively. These forces are described in this chapter, and although this is essentially a personal opinion relating to the UK, the

Table 1.1 Top mollusc producers (aquaculture and capture) (FAO, 1999)

		Tonnes × 1000		
Oysters		1992	1997	% change
China	*Crassostrea gigas*	752	2328	+210
Korea	*C. gigas*	253	218	−14
Japan	*C. gigas*	245	218	−11
USA	*C. virginica*	152	178	+17
France	*C. gigas*	134	147	+10
USA	*C. gigas*	33	33	
	World Total	1711	3279	+92
Clams/cockles				
China	*Tapes philippinarum*	576	1257	+118
USA	*Arctica islandica*	192	164	−5
Japan	*T. philippinarum*	59	40	−33
Korea	*T. philippinarum*	67	31	−55
China	*Anadara granosa*	56	130	+131
Malaysia	*Anadara* spp.	66	84	+27
Indonesia	*Anadara* spp.	46	43	−6
	World Total	2101	2780	+32
Scallops				
China	*Pecten yessoensis*	338	1001	+196
Japan	*Pecten yessoensis*	402	515	+28
USA	*Placopecten magellanicus*	108	48	−55
Canada	*P. magellanicus*	92	54	−42
	World Total	1061	1746	+64
Mussels				
China	Mytilidae	539	398	−28
Spain	*Mytilus edulis*	139	189	+36
Italy	*M. galloprovincialis*	105	124	+19
Netherlands	*M. edulis*	51	93	+82
Denmark	*M. edulis*	136	91	−33
Korea	Mytilidae	22	71	129
	World Total	1339	1363	+2

reader will recognise that many of them have influenced bivalve fisheries in many other parts of the world as well.

Positive forces affecting production

Legislation

British statute is governed by a fundamental common law right, which allows the public to take shellfish from the sea and all tidal waters except

Table 1.2 European production of oysters and mussels in 1997 (FAO, 1999)

	Tonnes × 1000	
	mussels	oysters
Spain	188.8	3.5
Italy	124.4	
Netherlands	93.2	1.2
Denmark	90.8	<0.01
France	64.6	149.7
Germany	22.3	<0.1
Greece	30.7	0.4
UK	32.1	1.7
Ireland	18.0	4.8
Turkey	8.4	1.5
Portugal	0.5	0.6
Total	677.5	163.7

Table 1.3 UK shellfish landings in 1997 (MAFF, 1998)

Crustacea	Tonnes × 1000	£ million
Nephrops	31.1	63.5
Crabs	22.5	24.3
Lobsters	1.5	13.1
Shrimps	1.0	1.0
Molluscs		
Scallops	18.5	27.6
Queens	5.6	2.0
Squids	1.6	4.0
Cockles	19.5	3.6
Mussels	19.0	3.1
Oysters	1.6	*1.9
Periwinkles	2.9	2.7
Other	12.2	8.2
Total	137.0	155.0

* includes data from Shellfish Association of Great Britain for 1100 t of farmed Pacific oysters with an estimated value of £1.1 million

where a private property right exists, or by restriction of a Parliamentary enactment. In Scotland this right excludes mussels and oysters, whose right of fishery is vested in the Crown. Such unrestricted freedom of bivalve harvesting, however, would be counterproductive, and legislation was introduced

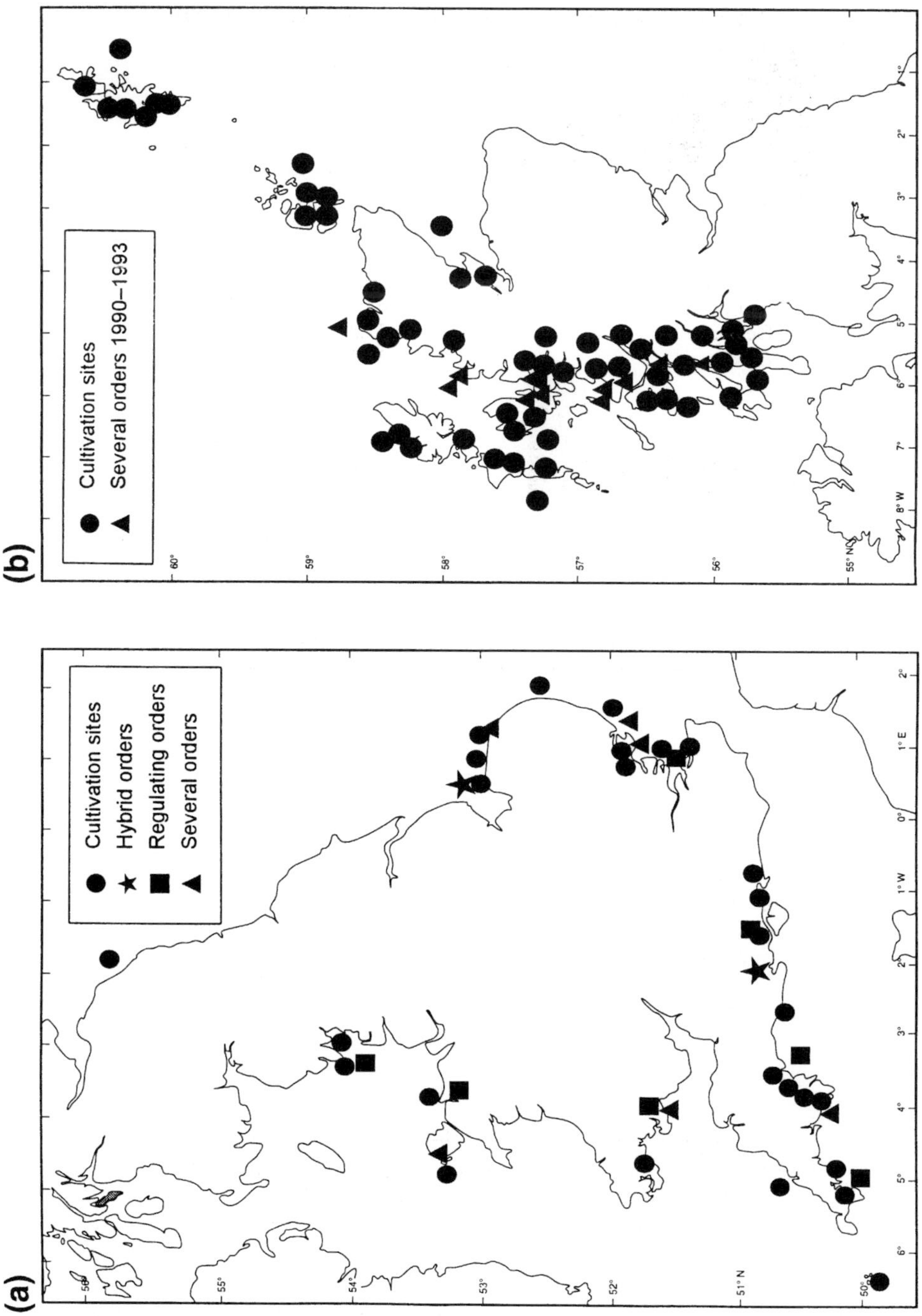

Fig. 1.4 Distribution of bivalve cultivation sites, (a) England and Wales, (b) Scotland.

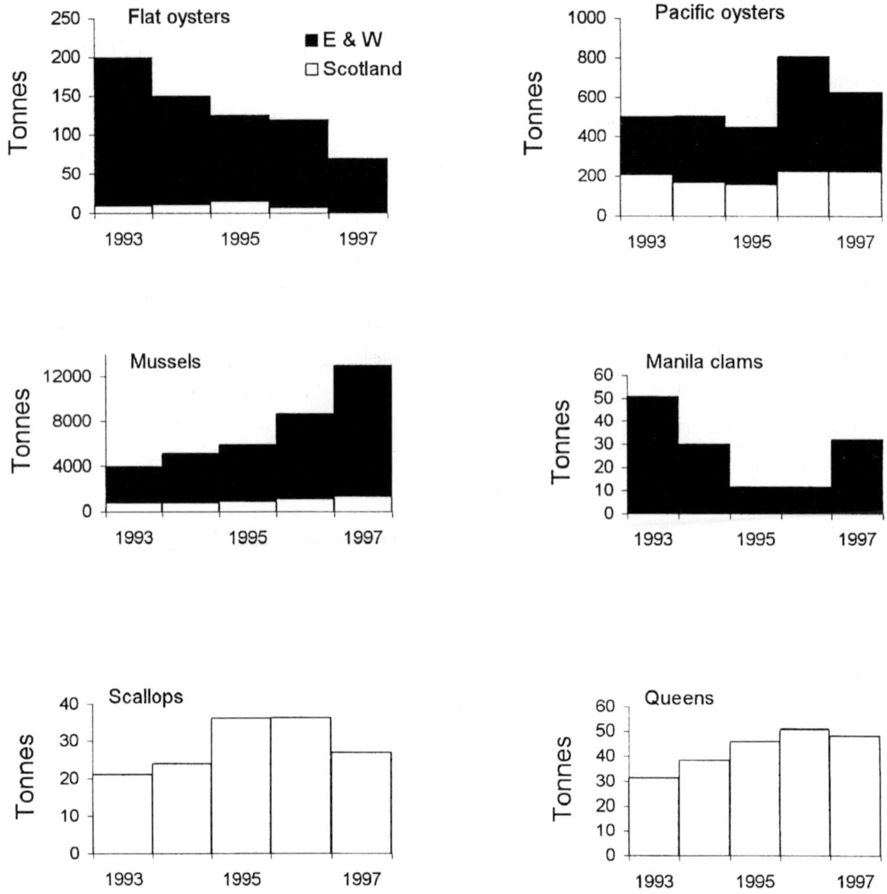

Fig. 1.5 Bivalve production from farms, 1993–1997 (England and Wales filled, Scotland unfilled symbols).

to ensure proper regulation of bivalve stocks by removing in some cases the public right to harvest bivalves.

Sea Fisheries (Shellfish) Act, 1967

There are many pieces of legislation that are important to the well-being of the bivalve industry. However, the Sea Fisheries (Shellfish) Act, 1967 has been one of the most important in structuring the present-day industry. It is the principal statute governing molluscan shellfisheries in Britain. Its origins lie back in the nineteenth century (1868) and the date shown here is its most recent update. It encompasses two bits of especially relevant legislation.

The Act introduced the concept of regulated and several fisheries, whereby the common law right of shellfishing is removed to ensure the protection of property rights in such fisheries. In 1998 there were seven regulating orders in England and Wales, protecting some of the major cockle, mussel and oyster fisheries. The total area involved was 143 800 ha, with individual areas ranging from 116 000 ha (Thames Estuary Cockle Fishery Order) to 156 ha (River Teign Mussel Fishery Order). There were also two hybrid orders [Poole Fishery Order (oysters, mussels, cockles and clams), and The Wash Fishery Order (oysters, mussels, cockles, clams, scallops and queens)], which are regulating orders with powers to grant leases of several rights. These add a further 72 000 ha to the total area. There were also 16 several fisheries in England and Wales with a total area of 2950 ha. These tend to be relatively small, ranging in size from 894 ha [Tollesbury and Mersea (Blackwater) Fishery Order (oysters)] to 50 ha [River Roach Fishery Order (oysters)]. The powers of granting several fishery rights have recently been introduced into Scotland. In 1990–2000, nine scallop several fisheries (two were for research and development purposes and one included native oysters), totalling *c.* 500 ha, were established in sea lochs, in support of the emerging importance of scallop cultivation there.

Regulated fisheries are managed by a set of by-laws which control, for example, gear specifications, minimum landing size, fishing season, quotas etc., while fishing effort is controlled by a licensing system. New by-laws may be introduced or old ones deleted, but only with government approval, as management requirements dictate. The several fishery concept, however, is designed to give the lessee a much greater management control of his stocks. The driving philosophy is that productivity of the fishery must be enhanced. Where natural stocks are concerned, enhancement may require encouraging spatfalls by the placement of collecting material (cultch, e.g. mussel, cockle or scallop shell), as is done periodically with flat oysters in the Solent, or the deposit of wild mussel seed on lays as in Poole Harbour. In Essex several fisheries and private grounds are traditionally stocked with half-grown flat oysters from the Solent for on-growing to market size. Some several fisheries may be used for the cultivation of non-native bivalve species, such as the Pacific oyster, grown intertidally in containers on trestles, and the Manila clam (*Tapes philippinarum*) grown in ground plots covered with plastic netting (Menai Strait and Poole Harbour).

Production of regulated fisheries may be quite variable from year to year because of inconsistent spatfalls or market demands.

Mussels

Mussel production from some of the regulated and several fisheries (Fig. 1.6) shows the irregularity of production over the past 30 years. The Wash has consistently been our main mussel producing area but annual landings have fluctuated between 10 000 t and <1000 t over this period, reaching an all-time low in the 1990s. The Conwy fishery has declined gradually owing to poor spatfalls, but has regained some lost ground by an active re-laying programme using mussels from exposed beds. Other fisheries, such as those in Poole Harbour and Menai Strait, have shown remarkable growth due to intense re-laying programmes using seed from adjacent or remote beds.

Cockles

Cockle landings from the main fisheries (Fig. 1.7) show a rise in the 1980s due to good spatfalls and increased market demand from the Netherlands where poor spatfalls have affected their fisheries. The Thames fishery remains our main source of supply with annual landings ranging between 1800 t and 25 000 t in the past 10 years.

Oysters

Flat oyster production has been low (less than 1000 t per year) since the 1930s. A further setback occurred in 1982 when the industry was severely hit by the disease *Bonamia*. Mortalities arising from the infection were high

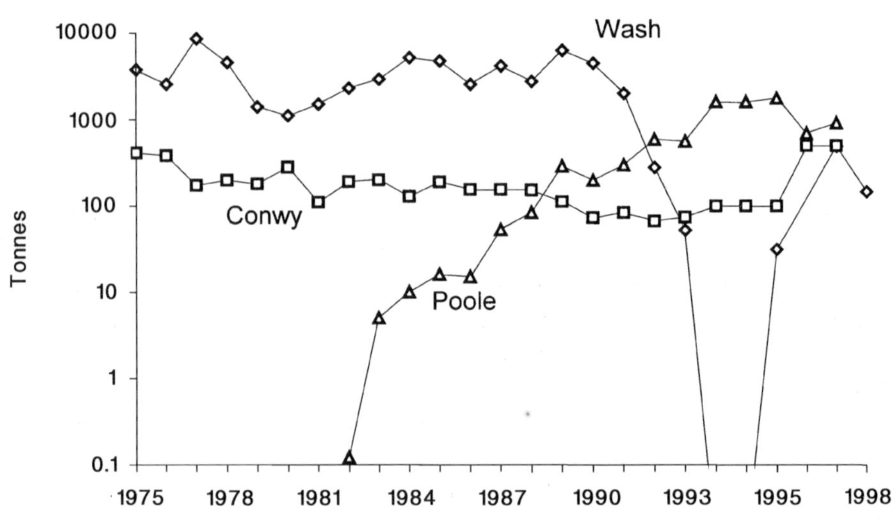

Fig. 1.6 Mussel production from some regulated and several fisheries.

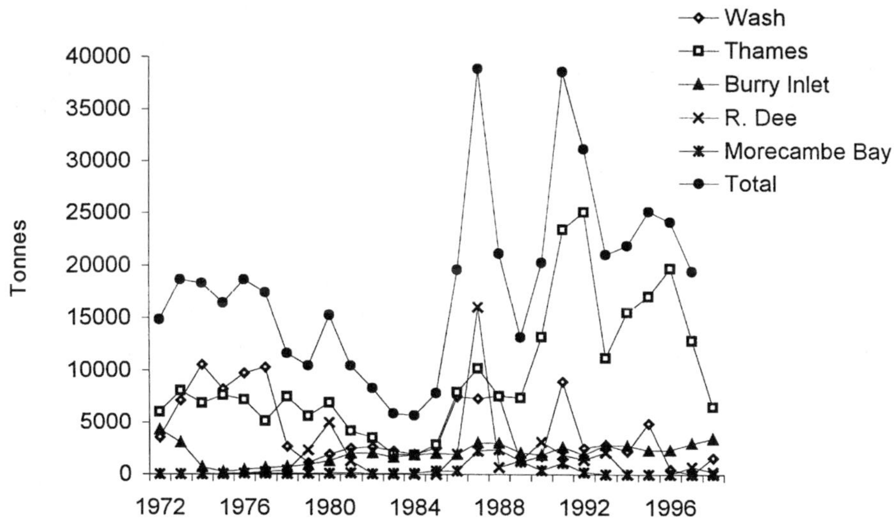

Fig. 1.7 Cockle landings in England and Wales.

on the natural and re-laying beds in the River Fal and also in the River Beaulieu, River Blackwater and Emsworth Harbour and at Walton-on-the Naze with re-laid oysters from the Fal and Solent. Nowadays, landings hover around 500 t/year (Fig. 1.8). They originate mainly from the regulated and several fisheries of the Solent, which remain relatively *Bonamia*-free, and from private and several fisheries in Essex using re-laid Solent oysters. The Loch Ryan fishery in Scotland has been producing up to about 20 t per year in recent years.

Molluscan Shellfish (Control of Deposit) Order, 1974; Variation Order 1983

The dwindling flat oyster fisheries at the end of the 19th century and the beginning of the 20th, received boosts from overseas with importations of unquarantined American oysters (*Crassostrea virginica*) up to 1939, and Portuguese oysters (*C. angulata*) from 1926 until the mid-1960s (Fig. 1.9). Up to 8 million oysters (300 t) per year were re-laid on British oyster beds for a growing season to fatten for the market. The outcome of these uncontrolled imports from America into parts of England and Wales was the introduction of two of our three main pests, the American tingle (*Urosalpinx cinerea*) and the American slipper limpet (*Crepidula fornicata*). The third pest, the 'red worm' (*Mytilicola intestinalis*), a gut parasite of bivalve molluscs, probably introduced from the Mediterranean, was first found in mussels at Southampton in 1937.

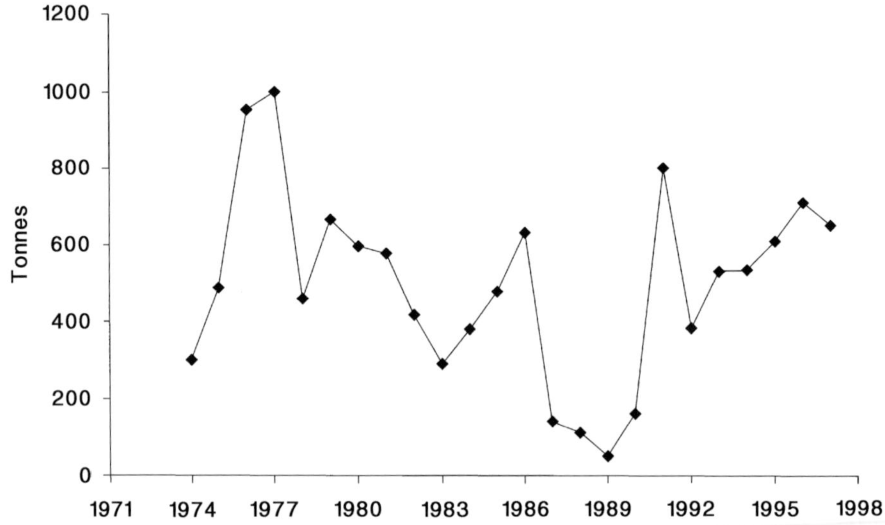

Fig. 1.8 Oyster landings in England and Wales.

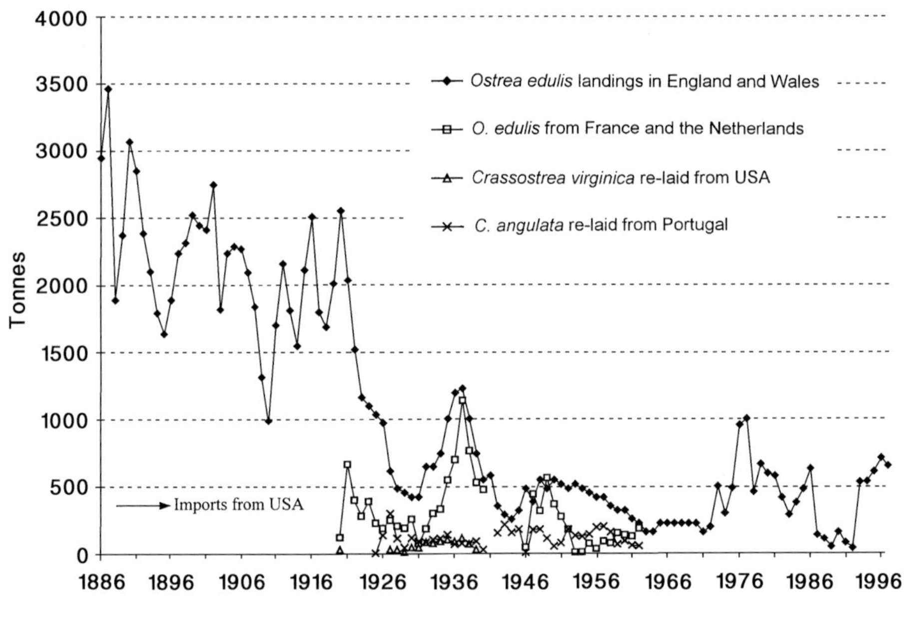

Fig. 1.9 Landings and imports of oysters in England and Wales since 1886.

The presence of these pests on cultivated bivalve grounds was generally considered to cause decreased production and higher working costs and it was recognised that to protect the long-term interests of the industry it was necessary to prevent the spread of these pests. There was a need, therefore, to control both imports of molluscs from overseas and also movements of molluscs around the country. Out of this need there arose the Molluscan Shellfish (Control of Deposit) Order, 1974. The Order essentially prohibited the deposit of molluscs from outside England and Wales (Scotland had similar controls) and the transfer of molluscs from one area to another, except under licence.

Since the order was first introduced in 1965 there was little evidence of the spread of *Urosalpinx* and *Mytilicola* and the spread of *Crepidula* by natural dispersion of larvae along the coast was slow. With the appearance of *Bonamia* in 1982, it was feared that the disease would run through the oyster fisheries. However, movement control by administratively segregating the areas infected, and good publicity, largely delayed the infection spreading to the Solent, although low levels of infection (*c.* 1%) existed on the beds at the end of the 1990s. The legislation, therefore, helped to control the spread of pests and disease in the UK. In 1991, a new EC directive (91/67/EEC) came into force, requiring less stringent conditions. The Fish Health Regulations, 1997, enabled this and a later directive (95/70/EEC, on the control of certain diseases affecting bivalve molluscs) to be introduced into UK legislation. This disregarded the pest status of *Urosalpinx*, *Crepidula* and *Mytilicola* and focused on the diseases *Bonamia* and *Marteilia*. The latter disease is not known to occur in UK waters. Only time will tell whether the new legislation will aid or hamper the shellfish industry; but already disquieting evidence of the establishment of *Mytilicola orientalis* in shellfish in the Republic of Ireland has occurred following the deposit of French Pacific oysters on some fisheries there.

Technology

Advances in gear technology in natural fisheries, especially that associated with processing, handling, harvesting (frequently originating from the continent) and depuration, have aided our shellfish industry in recent years. Cultivation of native species of bivalves has also benefited from imported technology. Collection of wild seed for suspended culture of mussels and scallops, using methods from Italy, Spain and Japan, adapted by the Sea Fish Industry Authority (SFIA) and the shellfish industry, are now making significant contributions to production in Scotland and Ireland and to a smaller extent in England.

Hatchery technology

UK hatchery technology is amongst the best in the world. Two methods of special significance, originating from Seasalter Shellfish hatchery, Kent, deserve a mention since they have transformed the face of hatchery technology, not only in the UK, but world-wide as well. The first method involved the production of cultchless spat. In the early days of hatchery technology, metamorphosing larvae of oysters were settled on a hard substrate, usually a matt-surfaced PVC disc. Good spatfalls, of many hundreds per cm^2, quickly caused competition for space and food, and as a consequence it required thinning to ensure good growth and survival. This itself caused high mortality. The solution was to remove the spat from the collector within 24 hours of settlement when they were *c.* 0.3 mm in size. This task could be achieved with a wide-bladed knife without causing too much damage to the spat. A more recent development has been to collect the spat on minute particles of oyster or other bivalve shell to achieve the same aim. These 'cultchless' spat were easier to handle but their true value has developed in conjunction with the second method of cultivation, which involves upwelling.

The upwelling method enables a deep bed of oysters or clams to be housed in a container so long as it is supplied with sufficient food-laden water to supply the nutritional and oxygen needs of the bivalves and to carry away waste products. This technology applies to small spat (<1 mm), recently settled in the hatchery, and to larger spat (3–15 mm) in outdoor nursery ponds supplied with pumped sea water or in floating upwelling systems in the sea, which rely on tidal exchange of water. The size of the container, mesh aperture and total water flow rate increase with size of bivalve and biomass stocked. Flow rates of around 20 ml/min/g of stock are adequate for good growth and survival.

All hatchery-reared oyster and clam seed produced in the UK, (around 100 million per year) are grown at some time of their early life history, in upwelling systems, showing the importance of this technological development to the UK industry.

Introduction of non-native species

Non-native species introductions have been of benefit to the UK shellfish industry. Seven species of bivalves have been introduced into the UK, under quarantine, via the Fisheries Laboratory, Conwy, since 1962 (Table 1.4). Not all of them proved suitable for cultivation and those that did not were destroyed. Some, however, showed enhanced growth and survival compared with native species and disease-free progeny of the imported parents were given to commercial hatcheries for large-scale production.

Table 1.4 Non-native bivalve species introduced into the UK under quarantine conditions

Year	Species	Common name	Origin	Fate
1962	*Tiostrea chilensis*	Chilean oyster	Chile	stock destroyed
1965	*Choromytilus choros*	Chilean mussel	"	" "
1963, 1965	*Tiostrea lutaria*	New Zealand oyster	New Zealand	self-sustaining population in Menai Strait
1965, 1972	*Crassostrea gigas*	Pacific oyster	Canada	commercially grown
	Crassostrea gigas	Pacific oyster	USA	commercially grown
1980	*Tapes philippinarum*	Manila clam	USA	commercially grown
1980	*Crassostrea rhizophorae*	Mangrove oyster	Brazil	stock destroyed
1984	*Crassostrea virginica*	American oyster	USA	broodstock given to industry

There are two reasons why new species were introduced. The first reason was to enable government scientific assessment of traits likely to be advantageous to the industry. These included the fast growth and good survival of the Pacific oyster (*Crassostrea gigas*); the similarity of appearance of the New Zealand oyster (*Tiostrea lutaria*) to our native flat oyster (*Ostrea edulis*) and for its short larval life which favours settlement in the vicinity of the parent stock; and the suitability of the American oyster (*Crassostrea virginica*) for culture in areas of low salinity or high silt load, its resistance to *Bonamia* and its greater tolerance to tributyl tin (TBT).

The second reason for introductions was commercial pressure. For example the Manila clam (*Tapes philippinarum*), introduced as a result of competition from French hatcheries, was considered to be a highly lucrative species which grows faster than our native species, the palourde (*Tapes decussatus*). The introduction of the mangrove oyster (*Crassostrea rhizophorae*) was requested by a University for research purposes, but with no potential for cultivation in our waters.

The value of the Pacific oyster to our industry today lies in the fact that its production (*c.* 600 t/year) is now higher than that of the native oyster from the natural fisheries. Manila clam production remains quite low at <50 t/year.

In 1973, a Working Group of the International Council for the Exploration of the Sea (ICES) produced guidelines and a code of practice for the '*Introduction of Non-indigenous Marine Species*' for member countries. This has been revised periodically, most recently in 1994, to take into account genetically modified organisms. The guidelines are:

- consider ecological consequences;
- review historical documentation of past introductions;
- seek advice from ICES on a particular introduction;
- adopt code of practice and report progress;
- develop legislation controlling introductions;
- propagate under quarantine in a hatchery;
- examine offspring for disease/parasites/pests before transfer to the sea.

Once a decision to introduce a non-native species has been made, well-managed hatchery quarantine facilities are essential for holding the brood-stock in captivity. Effluent water from the holding tanks must be sterilised prior to disposal and the broodstock destroyed once first-generation spat have been produced. It is also necessary for the F1 spat to remain in quarantine for up to a year, with regular histological checks, to ensure their continued health and freedom from pests and diseases. In this way, the risks of introducing pests and diseases not found in the UK are reduced to an acceptable level. Once these requirements have been met, the spat may be offered to commercial hatcheries as broodstock for spat production.

Probably the most contentious issue with introducing non-native species is associated with their breeding potential. For the past 30 years, it has been government policy not to introduce a species where there is a risk of it becoming established and creating self-sustaining populations. Although it is unlikely that Pacific oysters will become established to produce self-sustaining populations, research into sterilising non-native species of bivalves in the hatchery may provide a means of control. Genetic manipulation as a means of controlling reproduction has received considerable attention in recent years. This involves the production of sterile triploid offspring by inducing the eggs to retain three sets of chromosomes, rather than two, shortly after fertilisation. In the USA regular commercial production of triploid Pacific oysters is the norm, since the consequence of sterility is enhanced growth, meat content and flavour. Triploid oysters are now also produced by crossing tetraploid and normal (diploid) stock. Clearly these are traits for the shellfish industry to exploit. Triploid Manila clams, however, do not appear to offer the same advantages. The level of triploidy achieved within batches has not been greater than about 70% (>90% with Pacific oysters); there are no growth advantages and there is evidence that their survival is poor.

Pond technology

Pond technology for breeding flat oysters is a success story that needs to be recorded. Interest in developing the method began in the 1860s. The ponds were built near to high water of spring tides, where they could be filled

Fig. 1.10 Oyster ponds, Cork Harbour (Photo: David Ll Hugh-Jones, Atlantic Shellfish Ltd).

with good quality water and then isolated for a period of time during the normal breeding season of the flat oyster (usually May to July). Spat were collected on bundles of twigs, wooden hurdles, shells, slates and stones. Following enormous success in 1868, when 80 million spat were produced from the 80-acre (32 ha) ponds at Hayling Island, Hampshire, ponds were constructed elsewhere but with only intermittent success. Research into pond technology occupied scientists at the Fisheries Laboratory, Conwy, for around 40 years (until the early 1960s, and was reported on by Peter Walne (Walne, 1974) in his Buckland Lecture. The message was one of inconsistent results and research effort was abandoned in favour of the more controllable conditions of the hatchery.

A young entrepreneur, David Hugh-Jones (Atlantic Shellfish Ltd), who was keen to invest in a successful oyster cultivation enterprise, however, solved the problem. He saw the solution to year-to-year variability of successful spatfalls, in the construction of many ponds. Between 1971 and 1978, he built 22 ponds in Cork Harbour, Republic of Ireland (Fig. 1.10). Each pond, measuring 25 × 25 × 2.5 m deep (1 million litres capacity) was butyl-lined and inter-linked by underground drains to allow transfer of water to ensure better management control. After 250 pond-years of experience and with three-quarters of the ponds productive, the yield has been about 2 million 5-mm spat per pond, providing sufficient spat for a yield of around 100 t marketable oysters per year. Despite a setback in 1987 when *Bonamia* wiped out most of his stock, conditions have improved to allow him to pursue the method once again.

Depuration systems

Although these systems are a great technological benefit to the industry, their need could easily be classified as a constraint, since without sewage discharge into the marine environment there would be no need for depuration. The Fisheries Laboratory, Conwy, owed its existence to the sewage pollution in its river, since research into methods of depuration were first investigated there in the early 1920s. Consequently, the chlorine method of mussel purification became the standard in this country until UV sterilisation became the norm about 20 years ago. In fact the Conwy tanks were still operating successfully until 1994, when they were closed through lack of funding to raise their standards to that perceived to be necessary by new EC legislation. In the last few years SFIA has re-engineered UV depuration technology with modular designs suitable for various scales of operation, ranging in size from units holding a few tens of kilos to up to 0.75 t of bivalves (see Chapter 12). Their designs, which have also taken into account recycled use of natural and artificial sea water, and management requirements, are so reliable that the shellfish need only an occasional public health check to ensure that coliform bacteria are within specified 'safe' levels.

Constraints on the industry

Irregularity of natural spatfalls

Fluctuations in annual yields of our cockle, mussel and oyster fisheries are largely but not entirely due to good or poor spatfalls. Markets, both within and outside of the UK, also have an influence. Production of cockles has ranged from 8000 t to 40 000 t, mussels from 4000 t to 10 000 t and oysters from 60 t to 1000 t since 1970 (Fig. 1.11). Some areas, such as Morecambe Bay, produce quite regular spatfalls of mussels (5 out of 6 years) but these are largely lost in the first year owing to autumnal gales. The Wash mussel fishery suffers from irregular spatfalls. In the past 15 years, good spatfalls have been recorded only twice, in 1978 and 1986. This fishery has been our largest mussel producer, with landings between 5000 t and 10 000 t per year, but recent years have seen landings drop to their lowest figures because of poor spatfalls. Data for the past 100 years (Fig. 1.12) shows the frequency of spat-

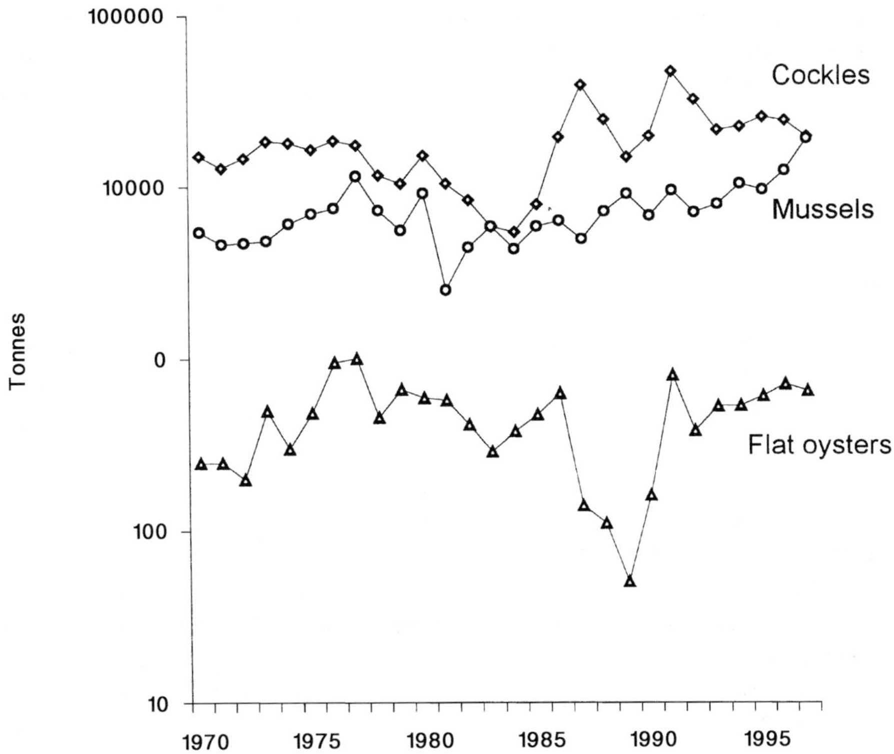

Fig. 1.11 Cockle, mussel and oyster landings in England and Wales since 1970.

falls in the Wash. On an arbitrary scale of 0 to 4, with 2 producing a sustainable fishery, spatfall levels are mostly below this value. Years with large spatfalls of a magnitude of 3 or more are infrequent and occur perhaps once every 5 or 6 years.

Factors that influence spatfalls are not well understood. Cold winters may be helpful by killing off shore crabs and reducing the predation pressure on the juvenile mussels. Stock biomass may also be important, not necessarily from large production of larvae, but from the large area that is provided for settlement. Since spatfalls are a natural event there is little that the cultivator can do to influence the course of events on the fisheries.

With oysters, traditional procedures for enhancing spatfalls include harrowing of the beds and laying cultch prior to settlement to provide clean surfaces of shells on which the larvae may settle. With such a poor record of spat production, the mussel industry must resort to the practice of re-laying from exposed to protected areas within the fishery, or from other fisheries which may be long distances away. In recent years, seed resources in

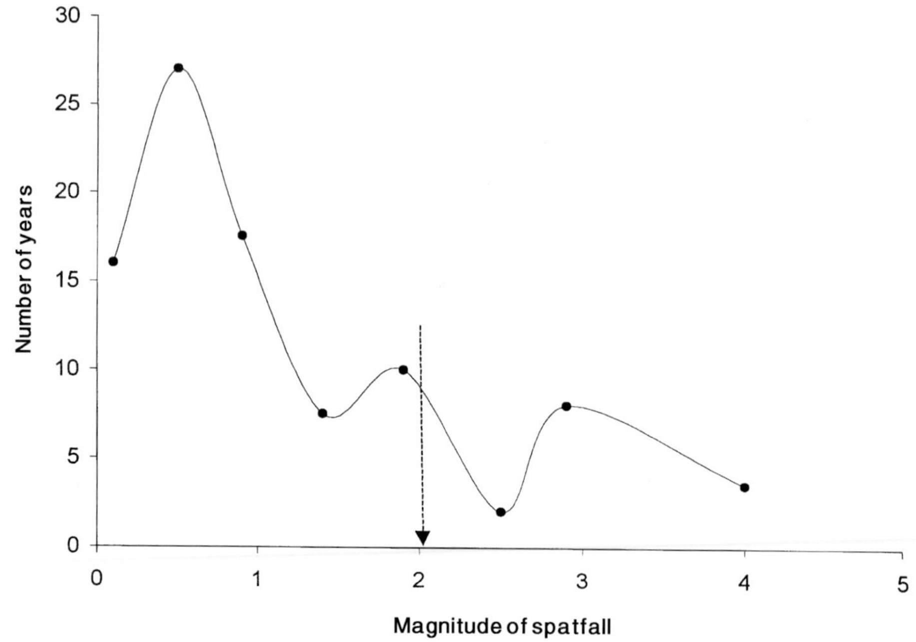

Fig. 1.12 Frequency of mussel spatfalls over a 100 year period in the Wash (Dare, 1993).

the Dornoch Firth, Portland Bill, Solway Firth and Morecambe Bay have been re-laid on various fisheries in England and Wales. A good example is provided by the growth of mussel cultivation in Poole Harbour, where cultivators have exploited the seed resource off Portland Bill.

Water quality

Sewage contamination

Sewage contamination of sea water has plagued the shellfish industry for over a century. Historical records show that public health has suffered periodically from the consumption of bivalves grown in waters receiving untreated sewage effluent contaminated with harmful bacteria and viruses. This problem was partially solved by the development of depuration processes that reduced the level of bacteria in contaminated bivalves to an acceptable level from the public health point of view. This highly expensive solution, which the shellfish industry has had to bear on behalf of the water authorities, is far from ideal. The more costly alternative of cleaning up sewage prior to discharge is an option that only now is beginning to emerge. It is essential that this should continue to enable the shellfish industry to

Table 1.5 Standards set for coliforms (*Escherichia coli*) in mollusc flesh

Category	Faecal coliforms (*E. coli*) per 100 g flesh	Comment
A	Less than 300 (<230 *E. coli*) in all samples	Suitable for consumption
B	Less than 6 000 (<4 600 *E. coli*) in 90% of samples	Depuration, heat or re-lay
C	Less than 60 000	Long re-lay (2 months)
D	Above 60 000	Prohibited

End-product standard (maximum permitted levels)
No *Salmonella* in 25 g flesh
PSP 80 µg/100 g flesh
DSP absence
ASP 20 µg/g flesh

Table 1.6 Classification of molluscan shellfish beds in 1999

	No. of beds classified	Percentage (%) of beds in classification			
		A	B	C	D
England and Wales	207	5	69	23	3
Scotland	161	43 (23*)	29	5	–

* A/B seasonal classification

create a healthier image. The Bathing Water Directive (76/160/EEC), the Shellfish Waters Directive (79/923EEC) and the Urban Waste Treatment Directive (91/271/EEC) should also help to improve the quality of estuarine and coastal waters following investment by water companies.

In the European Community, the Shellfish Hygiene Directive (91/492/EEC) required the classification of shellfish harvesting waters to ensure that a standard set of conditions are observed for the production and placing on the market of live bivalve molluscs. In the UK, the Directive was implemented by the Food Safety (Fishery Products and Live Shellfish) (Hygiene) Amendment Regulations, 1999. By 1999, 162 shellfish waters had been designated in the UK, with production areas classified according to the extent of contamination of the shellfish flesh with coliform bacteria (*Escherichia coli*) (Table 1.5).

The classification is revised annually, and the statistics for 1999 are shown in Table 1.6.

Shellfish poisoning

Paralytic shellfish poisoning (PSP), diarrhetic shellfish poisoning (DSP) and amnesic shellfish poisoning (ASP) are conditions of poisoning in people who have consumed bivalves containing toxin derived from dinoflagellates of the genera *Alexandrium*, *Gymnodinium* and *Pyrodinium* (PSP), *Dinophysis* and *Prorocentrum* (DSP) and *Pseudonitzschia* (ASP). PSP and ASP are potentially the most dangerous of the three since they have caused human deaths. Their distribution in the UK is mainly in the coastal waters around the north-east of England and north-east and west coast of Scotland. They occur mainly during the warmer months of the year, and although bivalves (mussels, oysters, clams, scallops, razor fish etc.) are mainly implicated, extensive testing has revealed that crustaceans feeding on them, e.g. brown crabs, swimming crabs and *Nephrops*, may also be affected.

Statutory monitoring of water samples and shellfish around our coasts (most intensively April–September) provides an early warning system. Short-term closure of beds in the event of toxins reaching critical levels has provided effective controls with minor economic losses to the industry and minimal threat to human health. In 1999 and 2000, however, ASP toxin severely affected the Scottish scallop fisheries, closing extensive areas offshore of the east and west coasts for prolonged periods, with an estimated loss to the fishery of *c.* £10 million.

A relatively new arrival on the scene is azaspiracid biotoxin (AZP), which although global in occurrence has been detected in Ireland, Scotland and Norway and has caused closures of beds in some areas. The toxin can cause severe gastroenteritis symptoms in humans but no fatalities are known. As detection methods become more routine and widely used, the causative organism and its distribution in space and time will become better understood.

Tributyl tin

Perhaps the greatest man-induced adverse effect on the oyster industry, apart from overfishing, is attributable to the use of anti-fouling paints containing tributyl tin (TBT) compounds on boat hulls. Historically, TBT-based paints were developed in the mid-1960s at about the same time as the Pacific oyster was introduced into UK waters. Abnormal shell thickening of the Pacific oyster in some areas, which produced 'golf-ball' shaped oysters with small meats, led to a survey of cultivation sites in the early 1970s. The prevailing opinion identified high suspended silt loads as the probable cause of abnormal growth of Pacific oysters. In France, Pacific oysters were also showing abnormal growth, with the degree of abnormality and the level of TBT in the flesh increasing with proximity to small-boat marinas.

Experimental work using plates coated with TBT-based paints showed that oyster mortalities occurred at concentrations of 2 μg/l organotin leachate and shell thickening at 0.2 μg/l. Subsequent work by government scientists in the

Blackwater and Crouch estuaries showed that TBT levels in the water frequently exceeded 0.2 μg/l, while laboratory-based work showed that growth of Pacific oyster spat was inhibited by exposure to 0.16 μg/l TBT and halted at 1.6 μg/l. In 1986, seasonal levels of TBT in the water at Burnham-on-Crouch showed levels exceeding the Environmental Quality Target (EQT) of 0.02 μg/l. Maxima of >0.12 μg/l occurred from April to November at times when newly painted boats were launched in the spring and when hulls were cleaned in the late summer prior to the annual regatta. Surveys of many of our shellfish cultivation sites recorded levels of TBT in the water, especially near to marinas, which were sufficiently high to explain the abnormal growth of oysters there.

Observations on flat oysters showed that although growth was not so dramatically affected as with Pacific oysters, there was a significant effect on their reproductive capability. Experiments showed that flat oysters subjected to TBT concentrations of 0.24 and 2.62 μg/l for a period of up to 74 days produced no larvae, whereas control oysters released *c.* 5 million larvae from an estimated 4–6 oysters of the 50 stocked. Also samples of wild oysters collected from within 0.5–1 km of dense boat moorings in the Crouch and Roach estuaries showed reduced larval production during the summer compared with those collected from a site offshore of the main estuary, away from TBT contamination.

Bivalves are amongst the most sensitive species to TBT. Other bivalves of commercial importance known to suffer severe growth retardation include the mussel (*Mytilus edulis*) and the palourde (*Tapes decussatus*), while the Manila clam (*Tapes philippinarum*) has proved to be quite resilient. Species of molluscs other than bivalves affected by TBT poisoning include the dog-whelk (*Nucella lapillus*) and offshore populations of the common whelk (*Buccinum undatum*) where they occur near to major shipping lanes. Rather fortuitously, the introduced American tingle (*Urosalpinx cinerea*), a major pest of flat oysters in Essex and Kent in the 1940s to 1960s, has also suffered severely from TBT poisoning. Populations of all of these gastropod species declined quite dramatically in estuaries and coastal areas subjected to TBT contamination. The decline was caused not by direct mortalities, but by the condition known as 'imposex' whereby the females developed male genital characteristics which effectively sterilised them.

The overwhelming evidence of the poisonous effect of TBT to a wide variety of marine organisms led to its ban in 1987 in paint formulations for use on boats <25 m in size and in antifoulants for use on fish farm cages. Nowadays, all paint formulations containing aquatic biocides require prior approval (Food and Environment Protection Act, 1985) before they can be used. Following the ban, water quality rapidly improved and Pacific oyster growth has now returned to normal.

Bonamia

The effects of this disease have already been discussed earlier in this chapter. Its origin in the UK is unknown but unconfirmed information suggested that it was transferred from France into the River Fal with unauthorised deposits of oysters. Although movement control was put into place fairly quickly, it occurred too late to prevent the spread of the disease. Diseased oysters re-laid from the Fal onto the beds in Essex, and from there to Poole Harbour and the River Beaulieu, threatened the natural beds in the Solent. An attempt to remove the diseased oysters from the Beaulieu failed, and after Emsworth Harbour became contaminated, it was only a matter of time before the Solent beds became infected.

There is no question that the effect of the disease on the flat oyster industry had been disastrous. The high level of the disease in the Fal prevented stocks from being used for relaying outside of the fishery. Fortunately, the low level of infection of Solent oysters (*c.* 1%) provided a source of half-grown oysters for re-laying in Essex. The local growers followed advice from the Fish Disease Laboratory (MAFF, later CEFAS), Weymouth, by re-laying infected stock at low densities, for one growing season and dredging and discarding small oysters to reduce the risk of carry-over of disease into the next season. In this way the Essex oystermen were able to pursue their business without undue mortality of their stocks.

Nature conservation

The Wildlife and Countryside Act, 1981 provided statutory means for controlling the release of non-native species of animals and plants into the wild. This caused some difficulties for the bivalve industry, for although the Pacific oyster and Manila clam were introduced prior to the Act, they could not be regarded as anything other than non-native species. MAFF, the then licensing authority, took the view that since Pacific oysters had been sufficiently long in cultivation in the UK before 1981, without any threat to the ecology of our coastal waters, they could be released under a general, unconditional licence. Since Manila clams were a relatively newer introduction, licensed releases were authorised only after consultation with the Nature Conservation Committee (NCC), or latterly the national conservation agencies Countryside Council for Wales (CCW), English Nature (EN) and Scottish Natural Heritage (SNH). Their concern centred on deposits being made within or near to Sites of Special Scientific Interest (SSSIs) or in enclosed areas where temperatures could reach sufficiently high levels that successful spawning might occur. However, since the seaward boundary of SSSIs in England and Wales extends only to mean low water and bivalve cultivation and their effects occur mainly below this level, discussions focused on ecological issues rather than any statutory considerations.

Possibly of greater consequence could be the European Union-inspired Birds Directive (EC 79/409/EEC) and the Habitats and Species Directive (EC 92/43/EEC) designed to protect the marine environment. Under these, Special Protection Areas (SPAs) and Special Areas of Conservation (SACs) respectively, are required to be designated by Member States, to create a European-wide network of protected areas. In the UK, many of the estuaries and coastal areas that have been designated, or are proposed for designation, are also sites where major bivalve fisheries or cultivation occur. Since the boundaries of SPAs and SACs can extend from high water mark to the limit of the Territorial Sea (or even beyond), there could be a conflict of interest with bivalve fishing or cultivation, especially in the use of intrusive harvesting methods.

Manila clam has had limited success as a cultivated species. Most commercial cultivation, small as it is, takes place in Poole Harbour (<50 t/year), in West Mersea (<2 t/year) and small experimental quantities in the River Exe. Searches in the Exe and Poole Harbour in 1990, jointly by MAFF and JNCC, did not reveal spatfalls of this species. However, following unusually warm summers spatfalls did occur in the early 1990s, leading to a small but significant commercial landing of Manila clams from Poole Harbour. Pacific oyster spatfalls also have occurred from time to time. Sitings were reported in Loch Sween and Emsworth Harbour in the early 1970s and more recently in the Rivers Exe, Teign and Dart in Devon and in the Menai Strait in Gwynedd. In earlier years, similar sitings of spatfalls of the Portuguese oyster were reported in the River Blackwater, but these were not self-perpetuating and eventually disappeared after imports were discontinued in the late 1960s.

To evaluate the impact of clam cultivation on the creatures living in the soil, the Fisheries Laboratory, Conwy, undertook field experiments in the River Exe using net-covered plots with and without clams. This study found that the netting acted as a substrate for the growth of macrophytic algae and encouraged suspended particles in the water to fall out of suspension. The sediment that built up beneath the netting was slightly enriched organically, presumably by the grazing activity of periwinkles, and housed higher densities of some suspension-feeding species of worms than those on unnetted plots. The use of disruptive suction devices for harvesting the clams can cause ugly scars on the intertidal beach, which visually heal quickly in high energy zones (e.g. cockle fisheries) and more slowly in low energy zones. In the Exe, a low energy area, it took 3–4 months for the site to return to visual normality and *c.* 18 months for the animal community to re-establish itself.

Although shellfish cultivation has been a feature of many of our estuaries for centuries, in some areas its presence has created some conflicts of interest with bird conservationists. In Scotland, suspended cultivation of mussels using floating longlines and rafts has provided a food source for small populations of eider duck. The wrangle that developed in controlling this problem was sorted out by the preparation of a code of practice by the Scottish Shellfish Growers Association, Scottish Natural Heritage and the Royal Society for the Protection of Birds, to minimise the impact to the mussels and to the birds. Shellfish farming activities on the foreshore is frequently cited as interfering with the feeding and roosting behaviour of waders, despite contrary scientific evidence. The circles of other legitimate estuary users, such as anglers, bait diggers, boat users and their moorings, etc., are increasingly beginning to overlap. This problem has been addressed in some areas e.g. Exe, Chichester Harbour and Salcombe estuary, with the preparation of management plans that attempt to bring order and control to the various activities. Clearly, dialogue between conflicting user groups is the right way to solve any differences that may exist between them with, as some estuary management plans recognise, the need to establish zones within which certain activities are confined.

Marketing

Marketing may be considered to be a constraint to the industry, since because of its small size and fragmented distribution, the onus of selling is often the burden that the grower has to bear if he is to survive. Small shellfish growers, however, may not be sufficiently skilled or motivated in sales promotion, which needs specialist input. Local markets are usually the first to be targeted, with pubs, restaurants and hotels providing outlets for locally-grown bivalves. As local markets become saturated, the search for new sales opportunities becomes increasingly wide until national and international markets are sought.

Crustaceans rather than bivalves appear to appeal to the public. The Sea Fish Industry Authority and the Shellfish Association of Great Britain have been involved in educating the British public into increasing their consumption of bivalve products over many years, but still the majority of our molluscan products are exported to the Continent of Europe. Greater home consumption of molluscan shellfish undoubtedly would stimulate a greater investment in the home industry.

References and Bibliography

Anon. (1993) Man's impact on ecosystems. In: *North Sea Quality Status Report, Oslo and Paris Commissions, London*. Olsen & Olsen, Fredensberg, Denmark.

Dare, P. J. (1993) Biological considerations in the exploitation of mussels. The Dr Walne Memorial Lecture. *Proceedings of the Shellfish Association of Great Britain*. Fishmongers Hall, London, England.

FAO (1999, 2001) *Yearbooks of Fisheries Statistics for 1997 and 2001.* **84**, **88** (1,2). Food and Agriculture Organisation of the United Nations, Rome, Italy.

Gibbs, P. E., Spencer, B. E. & Pascoe, P. L. (1991) The American oyster drill, *Urosalpinx cinerea* (Gastropoda): evidence of decline in an imposex-affected population (R. Blackwater, Essex). *Journal of the Marine Biological Association of the United Kingdom*, **71**, 827–38.

ICES (1995) *ICES Code of Practice on the Introductions and Transfers of Marine Organisms.* ICES Cooperative Report No. 204.

MAFF (1998) *UK Sea Fisheries Statistics for 1997.* Ministry of Agriculture, Fisheries and Food, Her Majesty's Stationary Office, London, England.

SOAFD (1998) *Scottish Shellfish Farms.* Annual Production Survey for 1997 (ed. D. I. Fraser). Scottish Office, Agriculture, Environment and Fisheries Department, Marine Laboratory, Aberdeen, Scotland.

Spencer, B. E. (1996) *Clam Cultivation: Localised Environmental Effects.* Ministry of Agriculture, Fisheries and Food, Directorate of Fisheries Research, Fisheries Laboratory, Conwy, Information Pamphlet.

Thain, J. E. & Waldock, M. J. (1986) The impact of tributyl tin (TBT) antifouling paints on molluscan fisheries. *Water Science and Technology*, **18**, 193–202.

Waldock, M. J., Thain, J. E. & Waite, M. E. (1987) The distribution and potential toxic effects of TBT in UK estruaries during 1986. *Applied Organometallic Chemistry*, **1**, 287–301.

CEFAS (1998) *Shellfish News* (ed. I. Laing). The Centre for Environment, Fisheries and Aquaculture Science, Lowestoft, England, **6**.

Walne, P. R. (1974) *Culture of Bivalve Molluscs – 50 years experience at Conwy.* Fishings News Books, Farnham, Surrey, England.

2 General biology of bivalves with respect to cultivation

Introduction

Molluscs belong to the phylum Mollusca, in which it is estimated that there are over 50 000 species. It is an ancient group of animals with the fossil record showing their presence 500 million years ago. It is also a diverse group with species occupying land, fresh water and marine habitats from abyssal depths to high intertidal zones and from polar to tropical seas. The phylum is divided into seven classes (Table 2.1), of which three, the Gastropoda, the Pelecypoda and the Cephalopoda, contain the majority of species (*c.* 88%, 9% and 1% respectively) and all the commercially exploited edible marine molluscs. Some edible molluscs are terrestrial, some are harvested from fresh waters, but the majority is harvested from the marine environment. All of the cultivated molluscs, apart from a small group of gastropods (haliotids or abalones), are bivalves and include the mussels, scallops, oysters, clams and cockles. For this reason this chapter is concerned with general aspects of the biology of bivalves, rather than gastropods, which are dealt with superficially later (see Chapter 9).

The evolutionary pathway of molluscs is speculative. It is assumed that the early marine ancestors evolved a protein matrix over their bodies, later consolidated by the addition of calcium carbonate to form a dome-shape shell. This provided a protective covering against predation, enclosing the soft organs between the umbrella of the shell and a muscular foot. Bivalves may well have evolved from this condition, before the advent of calcification, with lateral compression of the body and extension of the mantle and protective covering to enclose the body totally. The change in habit, from a directionally crawling mollusc to one with limited powers of movement, enclosed within a pair of shells, was accompanied by the loss of the head, a reduction in the importance of the muscular foot as a means of locomotion and a specialised ciliary feeding mechanism. Subsequent radiation of the bivalve form has been wide and varied, allowing the group to occupy a broad range of freshwater and marine habitats. It is assumed that the primitive bivalve burrowed in the sediment. It had two similar-sized adductor muscles (isomyarian condition) for shell closure, and had an inhalant

Table 2.1 Classification of the Mollusca, showing subdivisions of Classes in which cultivated molluscs are found

Phylum: MOLLUSCA				
Class CAUDOFOVEATA (Aplacophora; a small group of burrowing, worm-like molluscs with spicules instead of shell)				
Class SOLENOGASTRES (Aplacophora; a small group of surface-dwelling, worm-like molluscs with spicules instead of shell)				
Class POLYPLACOPHORA (chitons, with shell in a series of plates)				
Class GASTROPODA (univalve snails, e.g. whelks, periwinkles, abalones, limpets etc., and slugs with reduced or absent shells)				
	Subclass PROSOBRANCHIATA			
		Order ARCHAEOGASTROPODA		
			Superfamily HALIOTIDACEA	
				Family Haliotidae (abalone)
Class SCAPHOPODA (elephant tusk shells)				
Class PELECYPODA (Bivalvia or Lamellibranchia; with shell of two valves e.g. mussels, oysters, clams, cockles and scallops)				
		Order MYTILOIDA		
			Superfamily MYTILACEA	
				Family Mytilidae (mussels)
		Order OSTREOIDA		
			Superfamily OSTREACEA	
				Family Ostreidae (oysters)
			Suborder PECTININA	
			Superfamily PECTINACEA	
				Family Pectinidae (scallops)
		Order VENEROIDA		
			Superfamily CARDIACEA	
				Family Cardiidae (cockles)
			Superfamily VENERACEA	
				Family Veneridae (Venus clams; carpet shells)
Class CEPHALOPODA (squid, cuttlefish, octopus)				

water current entering the mantle cavity generally from around the ventral gape of the valves, and an exhalant current leaving via the posterior margin (Fig. 2.1). From this condition, two pathways have been suggested, with reduction in size of the anterior adductor muscle (heteromyarian condition), or its loss (monomyarian condition), changes to the control of the feeding current, and the retention of the byssus (essentially a larval condition) into adulthood. These changes have led to the specialised forms seen in modern-day clams, mussels, scallops and oysters.

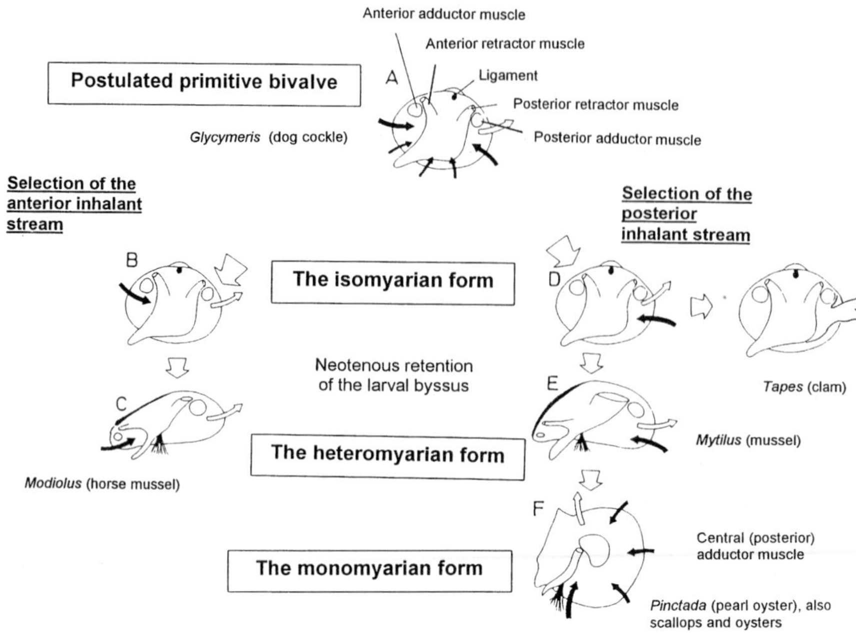

Fig. 2.1 Evolution of the heteromyarian and monomyarian forms from an isomyarian ancestor. (A) primitive anisomyarian bivalve; (B) selection of the anterior inhalant stream; (C) modioliform shell, e.g. horse mussel, *Modiolus* sp.; (D) selection of the posterior inhalant stream, leading to the heteromyarian form, e.g. mussel, *Mytilus* sp., and eventually to the monomyarian form, e.g. pearl oyster, *Pinctada* sp., scallop, *Pecten* sp. and oysters *Ostrea* and *Crassostrea* sp. (adapted from Morton, 1992).

Shell and mantle

The most obvious feature of the bivalve is the shell, consisting of two calcified valves joined together by an uncalcified elastic ligament. The outer fold of the edge of the mantle secretes the shell. The mantle also has a muscular inner fold that largely controls the flow of water entering and leaving the mantle cavity generated by the cilia on the gills, and a middle fold with a sensory function via the numerous tentacles and in some species, e.g. scallops, eyes.

During growth, the shape of the shell depends on the amount of marginal increase around the mantle rim. With unattached bivalves such as cockles and clams, there is symmetry around both ends, producing a rounded cockle shape or an oval venerupid shape. In some species, growth may be unequal at one end (i.e. usually the posterior end is enlarged), producing elongated shapes such as in mussels and razor shells. The transverse component of shell growth may result in globular-shaped cockles, flattened tellinids

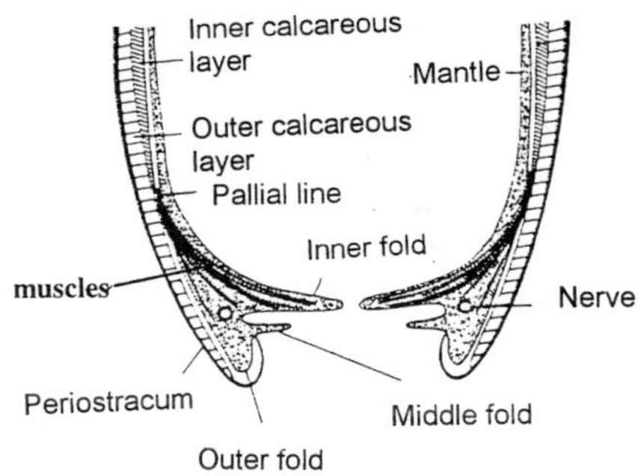

Fig. 2.2 Section through the shell and mantle of a bivalve showing the origin and nature of the shell layers and mantle folds (adapted from Yonge & Thompson, 1976).

and unequal valves, flattened on one side and rounded on the other as with oysters and scallops.

The shell comprises three layers, an outer horny periostracum and beneath, an outer and an inner calcified layer. The periostracum is a tanned protein formed by the cells on the inner side of the outer mantle fold (Fig. 2.2), extending to cover the outside of the shell. In oysters, this layer is thin and frequently wears down into a discontinuous layer or is lost. The outer surface of the fold produces the outer calcified layer, laid down within a matrix of protein (conchiolin) as prisms of calcium carbonate (either calcite or aragonite forms) on the inner surface of the periostracum. The inner calcified layer of the shell is produced by the general inner surface of the mantle, also within a protein matrix, and usually only within the pallial line. This is the only part of the shell that can be repaired following damage. The inner shell usually has the appearance of a dull glaze or is nacreous, in some species such as the pearl oysters (*Pinctada* and *Pteria*) forming the beautifully iridescent mother-of-pearl shells.

Inside the shell, the body is enclosed within the mantle. Although the general body form is similar between the various types of bivalves, major differences reflect adaptations to different modes of life. For example, the left and right lobes of the mantle may be separated around the rim in some species, e.g. scallops, allowing wide access for the inhalant water to enter the mantle cavity. In some species such as mussels, the mantle lobes are partially fused between a narrow inhalant and a wide exhalant siphon, while in clams the mantle lobes are joined to form elongated siphons allowing the animal to burrow beneath the substrate yet retaining contact with the water above. Also, the foot is usually well developed in burrowing forms such as the clams, reduced in size in byssally attached forms where it acts as a plantar during attachment of the byssus to the substrate, or absent in those species, e.g. oysters, that cement themselves to the substrate.

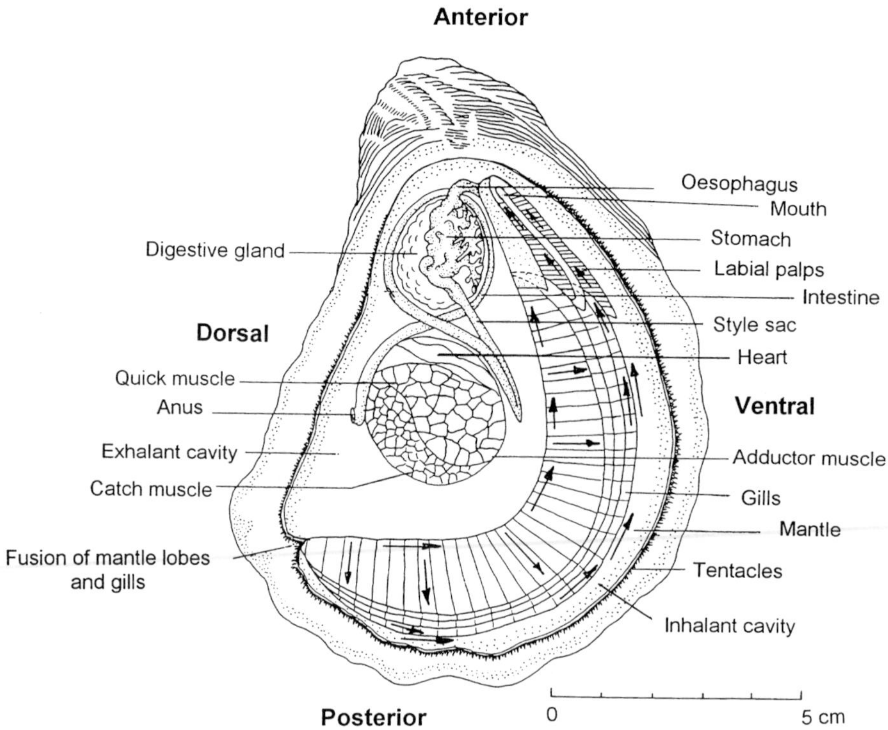

Fig. 2.3 Cupped oyster with right valve and mantle removed, showing general disposition of gut and gills around central adductor muscle (adapted from Kennedy *et al.*, 1996). Arrows indicate direction of movement towards mouth of particles trapped on gills.

In the monomyarian bivalves (oysters and scallops), the body organs are arranged around the centrally placed adductor muscle (Fig. 2.3). Shell closure is controlled by the contraction of the adductor muscles working against the resistance of the ligament, which becomes compressed during closure. Weak or dead bivalves gape owing to the uncompressed state of the ligament. The adductor muscles of most types of bivalves consist of two kinds of muscle, which are either slow- or fast-acting. The quick part of the adductor is located on the side nearer to the hinge (Fig. 2.3). It is translucent in appearance and is responsible for the rapid closure of the shell in response to disturbance, to clear accumulated pseudofaecal waste from the mantle cavity or for swimming activity in bivalves such as scallops. The slow part of the adductor is opaque, acting as a 'catch' in those species that remain closed for prolonged periods.

The largest organs are the paired ctenidia (gills), signifying their importance for food collecting. The attachment of the gills to the mantle at the anterior, posterior and dorsal edges, effectively divides the mantle cavity into inhalant and exhalant compartments. Water passes from one compartment to the other, through the pore structure of the gills, where food particles are removed by ciliary action. In *Crassostrea* species, water may freely flow from the inhalant to the exhalant chamber through the promyal chamber, an adaptation enabling the oysters to inhabit turbid waters. Food is carried on ciliated tracts to the mouth at the anterior end of the labial palps. The gills are also well supplied with blood vessels and are the major site for respiration.

The gill filter feeding mechanism of commercially exploited bivalves conforms to the general lamellibranch pattern. It is driven by a well-organized array of cilia and tracts that controls water flow into and out of the mantle cavity and entraps and directs food particles in suspension towards the mouth or away for rejection. The dependency of lamellibranchs on particles, largely phytoplankton as small as 1 μm in diameter, as food has created a pair of enlarged gill structures occupying much of the mantle cavity. Elongation and reflection of the filaments into ascending and descending arms (demibranchs) on either side of an axis forming a W-shaped fold (lamella) (Fig. 2.4A) have achieved further increase in size.

In the more primitive filibranch gill (e.g. mussels and scallops), the filaments are similar in structure and are joined laterally by ciliated tufts (interlamellar junctions). In the majority of lamellibranchs, e.g. oysters, clams and cockles, the gill is a more complex structure, with further increase in surface area achieved by the development of folds and crests (plicae) (Fig. 2.4B). Each plica comprises several types of filament (ordinary, transitional and principal filament), each joined by permanent tracts of vascular tissue.

In the plicate gill, water is driven from the inhalent to the exhalent cavity via the numerous small openings (ostia) by lateral cilia arranged along the sides of the filaments (Fig. 2.4B). Rows of latero-frontal cilia, which spread across the opening between filaments, filter the water and transfer particles onto the apex of the filament.

Current theory suggests that filtration is achieved hydromechanically, with a complex three-dimensional flow past the gill filaments created by the latero-frontal and frontal cirri. These produce steep shear gradients in water velocity, which draw particles to the gill surface for entrapment. Once trapped on the filaments, the food particles are transported by the frontal cirri to ciliated tracts in the ventral marginal groove, bound in mucous or dorsally to the basal groove, and then onwards loosely bound as a slurry towards the mouth. Prior to ingestion, the food particles from the two food grooves are processed by the labial palps (Fig. 2.5). The particles in the mucous string are disaggregated by mechanical action of the palps. Together with the particles from the basal food groove, they are transferred to the

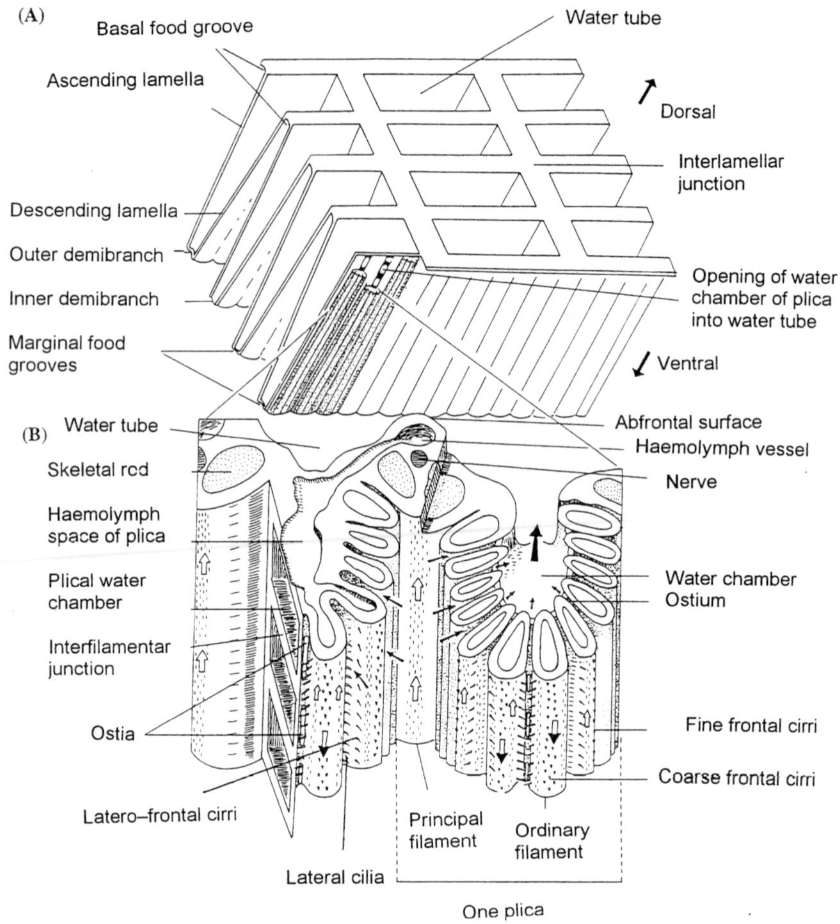

Fig. 2.4 Diagram of (A) part of two gills of *Crassostrea virginica*, and (B) detail of plical fold. Direction of movement of particles on frontal cilia is indicated by black headed and open arrows. Direction of water movement between the filaments, through the ostia and into the water tubes is indicated by the solid black arrows (adapted from Kennedy *et al.*, 1996).

mouth at the anterior end of the palps or to the ventral margin and onto the mantle for disposal as pseudofaeces (Fig. 2.5). In some conditions of high particle loads in the water, the sorting process may be suspended and mucous strings rejected as pseudofaeces (Fig. 2.5A).

Digestion

Ciliated tracts carry food particles entering the mouth to the stomach where the first stages of digestion take place. In oysters, the stomach is divided into

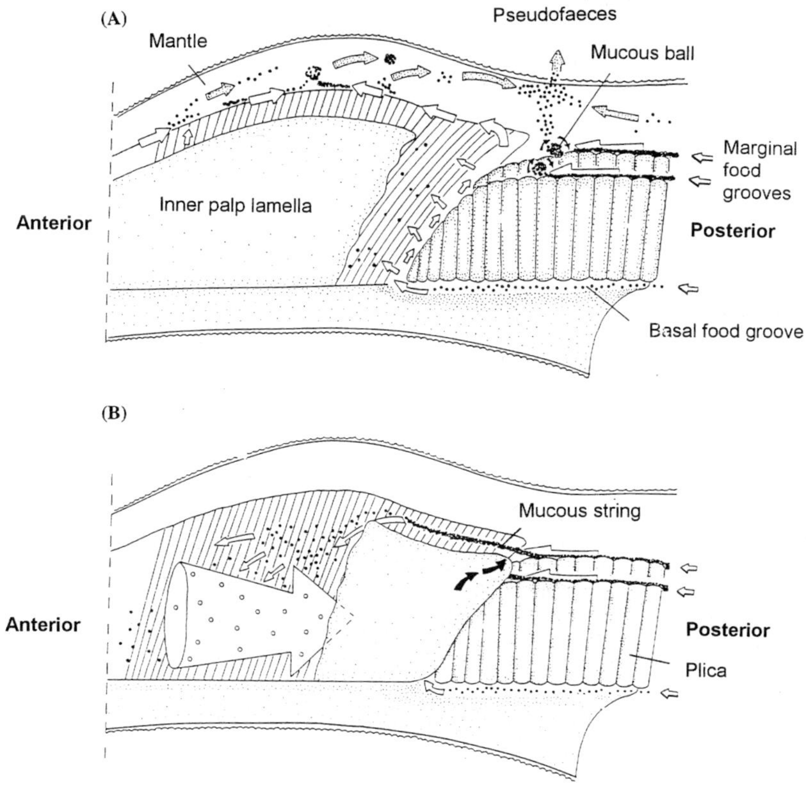

Fig. 2.5 Lateral view of gills and labial palps of *Crassostrea virginica*. The inner labial palp lamella is drawn part removed to show ridged surface of opposing lamella. Arrows show direction of particle movement. (A) The palps are withdrawn away from the gill margins, with mucous balls forming at termini of marginal food grooves for disposal as pseudofaeces on mantle rejection tracts (stippled arrows). Transfer of particles from the basal food groove continues. (B) Palps in contact with gills receive mucous strings of particles for disaggregation and sorting (solid circles) for ingestion or rejection (from Ward *et al.*, 1994).

two chambers in which the particles are sorted by size on ciliated ridges and grooves and mixed with digestive enzymes. Further mixing and digestion occurs in the posterior chamber, specifically by the crystalline style, an organ unique to the molluscs, having evolved separately in some gastropods and bivalves that feed herbivorously on small plants or phytoplankton. The style is a flexible, gelatinous rod housed in a blind sac that impinges onto a cuticular structure on the wall of the posterior chamber of the stomach known as the gastric shield. The action of ciliated tracts on the walls of the style sac causes the rod to rotate against the shield, breaking down algae cells and releasing digestive enzymes from the rod and the gastric shield. Although the head of the rod dissolves during this process, it is continuously reconstituted in the sac. In some species of bivalves including oysters, in which the style sac is not completely separated from the intestine, the rod may vary in size depending on environmental and feeding conditions and usually dissolves when the bivalve is out of water.

Small and partially digested particles in the stomach are carried to the vicinity of the openings of the digestive gland and are drawn into the tubules of this brown-coloured organ surrounding the stomach. The food particles are phagocytosed (engulfed) by specialised cells in the tubules and incorporated into vesicles containing digestive enzymes where intracellular digestion takes place. Undigested particles are returned to the stomach, mixed with other unwanted matter and transferred to the lower intestine. During its passage down the intestine, the food is compacted into ribbon-shaped faeces. These are shed into the exhalant cavity for disposal with the outgoing flow of water.

Circulatory system

Bivalves have a simple, open type of circulatory system. The heart pumps blood haemolymph through arteries that branch throughout the body and open into sinuses where the organs are bathed via intercellular spaces. Blood from the organs is returned to the heart via the gills and mantle, where it is re-oxygenated. The blood haemolymph carries cellular haemocytes around the body. There are several types of haemocytes (granulocytes, halinocytes and serous cells) of which the first two are capable of phagocytosing particles with their finger-like pseudopodia. Oyster haemocytes are involved in a range of vital functions including wound and shell repair, food digestion and transport, gaseous exchange in respiration, excretion and internal defence. Food particles digested within haemocytes may be carried from the digestive gland or stomach to deep within the tissues, while indigestible waste materials are voided by migration of phagocytes across the epithelial borders of and into the alimentary canal for disposal outside of the body.

Bivalves may accumulate high levels of heavy metals in naturally or industrially polluted waters. Of particular concern are copper, zinc, mercury and cadmium, which may cause poisoning to the consumer if sufficiently contaminated. The heavy metals are accumulated in the vesicles of the haemocytes, and in oysters tend to be stored in the gills, digestive gland and mantle without apparent harm. The greening of the gills of flat oysters in the River Fal, England, caused by the high levels of copper from past mining activities, and with the copper content reaching 1–2% of dry tissue weight, has led to unpalatable or poisonous oysters. It is speculated that accumulation of copper and zinc may play a part in defence against infection in oysters, since cells rich in these metals have been observed to migrate to sites of tissue damage where their release has a bacteriocidal effect on invading organisms.

Reproduction

Most bivalves have separate sexes (dioecious), a few are hermaphrodite, while unusually some are capable of alternating between sexes. Some clams, e.g. the quahogs, *Mercenaria mercenaria* and *M. campechiensis*, are consecutive hermaphrodites, exhibiting bisexual gonads as juveniles but changing to one or the other sex after the first year and remaining unchanged thereafter. Other bivalves, e.g. the scallop *Pecten maximus* and the giant clam *Tridacna gigas*, are functional hermaphrodites with the gonad containing well developed male and female sex cells. These are not usually shed simultaneously when ripe, thus minimising the risk of self-fertilisation. Oysters are alternate hermaphrodites, maturing first as males (protandry) then changing to femaleness and back again in a cycle that may be an annual event with the cupped oysters, *Crassostrea* species, or several times each year with the flat oyster *Ostrea edulis*. Environmental conditions, especially food and temperature, may determine the frequency of change and the outcome of the sex. There is some evidence that femaleness in *Crassostrea* species is the most likely outcome in times of food abundance. This is perhaps exemplified by the observation that small oysters in a population of old *C. gigas* kept in trays in the Menai Strait are usually male while the larger oysters are female.

The reproductive system of bivalves is simple, comprising a pair of gonads that open into the exhalant cavity. The tissue is little more than a complex of ramifying tubules. In mussels, the gonad is located in the mantle; in oysters it is interspersed within the visceral mass eventually covering it as it thickens and ripens; and in clams it extends into the foot. With ripening, the gonad may develop distinctive colours. In oysters, it is a creamy white colour irrespective of sex. In mussels, the male gonad is usually creamy white and the female an orange colour. Perhaps the most striking difference occurs in the gonad of the hermaphrodite scallop with its sharp demarcation in colour between the creamy white testis and the deep orange ovary.

When ripe, most bivalves shed their sperm and eggs into the exhalant cavity and then to the outside with the outgoing current, where fertilisation occurs. In oysters, however, although the sperm passes to the outside the eggs do not. They pass through the ostia of the gills, which become enlarged owing to the relaxation of their muscles, into the inhalant cavity. In *Crassostrea* spp. the eggs pass out of the inhalant cavity and are fertilised

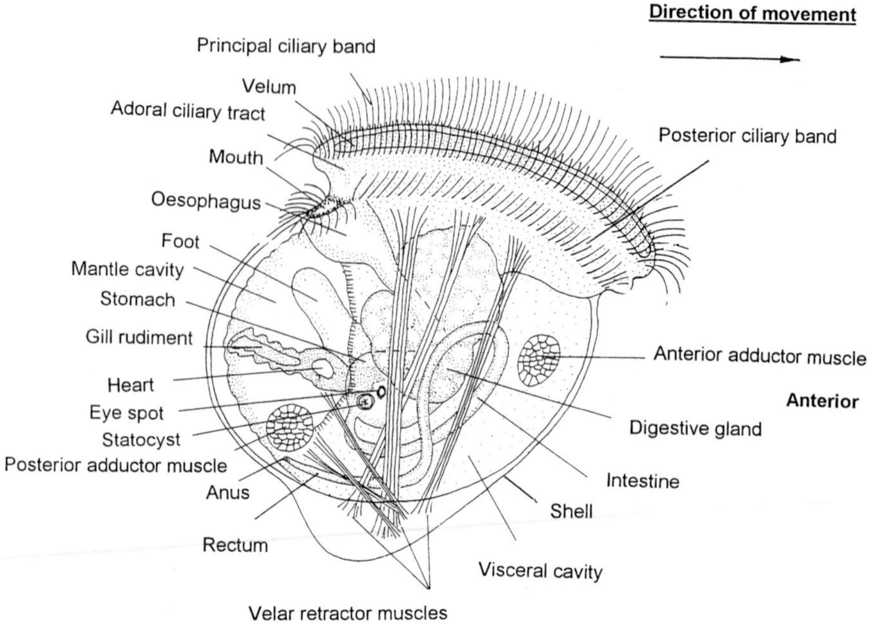

Fig. 2.6 Late veliger larva (shell length *c.* 250 µm) of *Crassostrea virginica* in swimming position (adapted from Newell & Langdon, 1996).

by sperm released by other oysters. In flat oysters (*Ostrea* and *Tiostrea* spp.) the eggs, fertilised by sperm drawn into the cavity, are retained and incubated for several weeks. In *O. edulis*, the eggs develop to a shelled veliger stage, changing in colour from white through grey to black as the shell develops and become pigmented. Fishermen call these stages 'whitesick', 'greysick' and 'blacksick', when seen as an oozing mass in opened oysters. The larvae (*c.* 170 µm in size), released from the parent approximately half way through the pelagic cycle, are puffed out of the cavity by the clapping of the valves. Once released, the larvae are capable of swimming and feeding by means of the ciliated currents generated by the velum (Fig. 2.6). Larval development continues for a further 1–2 weeks, when the pelagic phase ends (at >270 µm in size) with metamorphosis into the sedentary spat.

The process is similar in the New Zealand oyster, *Tiostrea lutaria*, except that the larvae are incubated for a longer period of *c.* 4 weeks and on release (>450 µm in size) are equipped with an 'eye spot' and foot and are ready for almost immediate settlement. Incubation appears to be a protective strategy evolved to optimise survival of larvae, since eggs removed from the parent are capable of normal growth and survival when cultured in the laboratory. To undergo prolonged incubation, the eggs require an energy reserve of which lipids are important. This requirement is reflected in egg size and fecundity. Eggs of non-incubators, e.g. mussels, clams, scallops and *Crassostrea spp.*, are small (50–80 μm) and require a lipid reserve to sustain them for a few days until they are able to feed. The eggs of *O. edulis* and *T. lutaria* are larger, *c.* 120 μm and 280 μm in size, respectively, and contain sufficient lipid to sustain them for much longer periods.

The fecundity of oyster species, size for size, is proportional to egg size. *Crassostrea* spp. produce 10–100 million eggs per female (usually spawned over a prolonged period of the summer), while *O. edulis* and *T. lutaria* produce around 100 000–1 500 000 and 10 000–120 000 larvae per female, respectively.

Lipid is also an important energy reserve during metamorphosis. It is accumulated during larval development in those species that feed during the free-swimming part of the larval cycle. The larvae of *O. edulis* increase their lipid reserves from *c.* 15% to 22% of the dry tissue weight during the pelagic phase, utilising this energy store during the few days of metamorphosis when they are incapable of feeding. The dependency on lipid as the main energy reserve of bivalve larvae lasts only for a few weeks into the spat stage when there is a switch to carbohydrate (glycogen) storage. This becomes the principal energy reserve for rest of its life.

The structure of bivalve larvae is fairly similar between the different groups, conforming to the same general plan (Fig. 2.6). With approaching metamorphosis, an eye spot (except in clams) and an exploratory foot develop as aids to selection of a site suitable for settlement. Once selected, the larva attaches by cementing itself (oysters) or by anchoring with byssus threads (clams, mussels and scallops) to its chosen site and undergoes metamorphosis. The cement or byssus gland is situated at the base of the foot and is instrumental in positioning the attachment element. The cement gland is lost after metamorphosis, but the byssus gland is retained for a while in scallops and clams and into adulthood in mussels. Within 2–3 days, other internal organs are modified to suit the new sedentary existence. The velum and eye spots disappear and the mouth moves through 90°. In the monomyarians, the anterior muscle disappears, with the posterior adductor muscle moving to a central position, while the foot disappears completely in oysters and is reduced to a minor role in scallops. At the same time, the first gill filaments develop, gradually increasing in number and complexity, enabling the juvenile bivalve to draw in the large quantities of water and food to sustain the rapid growth necessary for survival through the early stages of its life.

Food and feeding in cultivation

The type and quantity of food that a bivalve receives is of paramount importance to its welfare throughout all stages of the cultivation cycle. In general, the bivalve gill is capable of filtering particles efficiently down to *c.* 7 µm (Fig. 2.7). Efficiency may vary between and within species, the latter in response to environmental conditions such as particle load. Regulation of retention of particles may be a function of the beat of the latero-frontal cilia and/or by control of the pumping rate of water through the gills, or by muscular control of ostia pore size. In cultivation, live microalgae within the size range of 2–10 µm are the most commonly used source of food. Some effort, however, has been made in the use of dried algae and artificial diets, but in only some cases have they achieved success as part replacements to a natural diet.

Many species of microalgae have high food value while others have little or none. This is largely due to their nutritional value, although some are excluded owing to their large size or to the indigestibility of the cell wall. Microalgae vary widely in protein, carbohydrate and lipid content, and differences, especially in carbohydrate and lipid, can influence the nutritional characteristics of the diet. The fatty acid composition, especially the presence of long-chain components [polyunsaturated fatty acids (pufas)] such as 20:5n-3 and 22:6n-3, are known to influence the food value of microalgae, provided that other essential components are also available. Microalgae

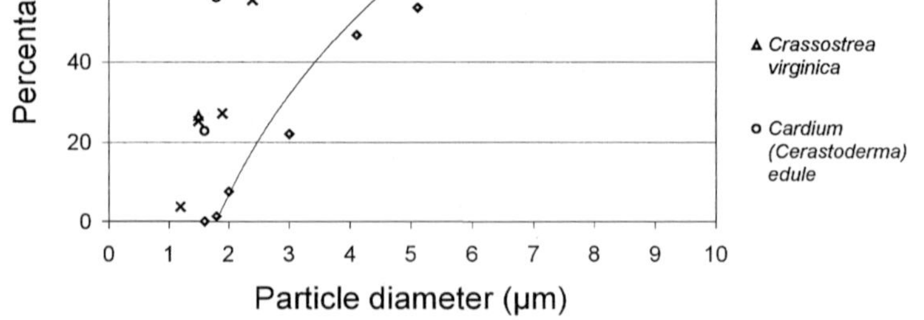

Fig. 2.7 Particle retention efficiency of various bivalves (redrawn from Vahl, 1973). Trend lines for *Mytilus edulis* and *Chlamys opercularis* are shown.

lacking or rich in pufas are usually poor or good foods respectively. Species of good food value used extensively in aquaculture include *Chaetoceros calcitrans*, *Skeletonema costatum*, *Thalassiosira pseudonana*, *Isochrysis* sp (var. T *iso*) and *Pavlova lutheri*, while *Tetraselmis suecica* is of moderate value. Poor foods include *Dunaliella tertiolecta* and *Phaeodactylum tricornutum*. Although the latter species is rich in 20:5n-3 its spiculous silica cell wall is believed to reduce its food value.

Although single species diets may provide good growth of bivalve larvae and spat, better nutritional balance is obtained by feeding a mixture of species (see Chapter 3 on hatchery rearing).

Provision of the right quantity of cells is equally important as providing good food species. Adult oysters, clams and scallops kept as broodstock lose condition (flesh weight) with prolonged confinement unless fed adequate maintenance rations of microalgae with good food value. This amounts to *c.* 3% (dry algae per unit of initial dry flesh weight) for the scallop *Pecten maximus* and 6% for the oysters *Ostrea edulis* and *Crassostrea gigas* and clams *Tapes decussatus*, *T. philippinarum* and *Mercenaria mercenaria*). Correct feeding ensures higher egg yield, with the eggs rich in long-chain pufas, which in turn enhances larvae growth and spat production. Larvae also require the correct blend of algae species (see Chapter 3). Small larvae such as those of *C. gigas* (initially *c.* 70 μm) can only ingest small algae, e.g. *Chaetoceros calcitrans* (2 μm), but as they grow to a shell length of >100 μm they are capable of ingesting larger algae such as *Isochrysis* (5 μm) and *Tetraselmis* (10 μm).

Factors other than food quality and quantity play an important part in controlling food intake. In the cultivation process, it is especially important to ensure that growth and survival are kept adequately high by manipulating the biotic and abiotic factors that control them. Bivalve filtration is influenced largely by water flow rate, food concentration and temperature. These factors can be controlled in the early months of cultivation but less so as the cultivation process proceeds. For example, cultured microalgae provide the nutritional needs of the broodstock, larvae and early spat. With increase in size of spat, and therefore bulk, there is a greater dependence on naturally-produced microalgae. These needs may be met by adding fertilisers to outdoor nursery ponds to promote natural phytoplankton production. Extensive observations on factors influencing the filtration of food particles by a number of juvenile oyster and clam species in such systems have provided a useful insight into their management.

Increasing specific flow rate (flow relative to the weight of animals stocked), size and particle concentration causes a lowering of filtration while increasing temperature causes an increase. The relationship between these factors and filtration may be described by mathematical equations. Such an equation has been used to show how filtration declines with increasing flow

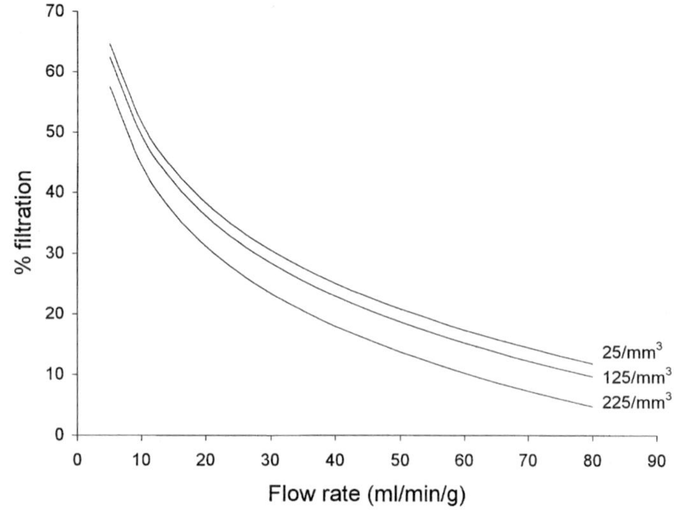

Fig. 2.8 Filtration by *Crassostrea gigas* in relation to water flow rate in upwelling tubes. Calculated for 10 mg spat at three levels of particle concentration and a temperature of 15°C [from % filtration = 74.4 − 19.03 ln V − 0.044 W − 0.086 C + 1.42 T; where V is specific flow rate of water (ml/min/g), W is mean live weight of spat (mg), C is particle concentration (number/mm^3), T is temperature (°C)] (from Spencer, 1988).

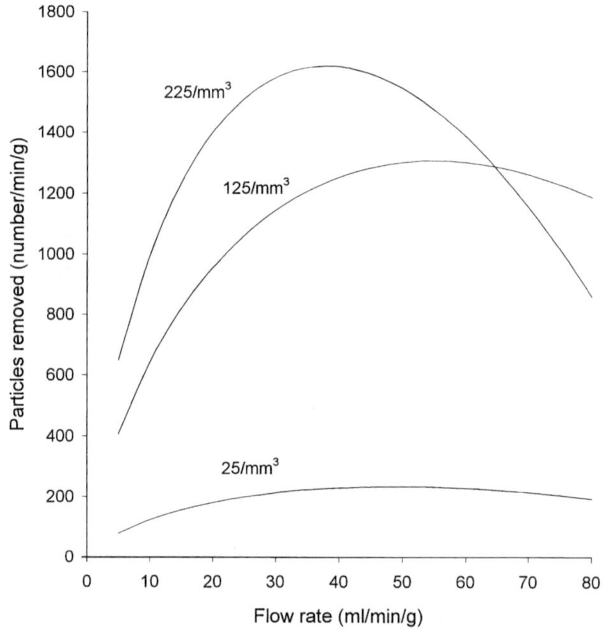

Fig. 2.9 Rate of particle removal by *Crassostrea gigas* in relation to water flow rate in upwelling tubes. Calculated from equation shown in Fig. 2.8 (from Spencer, 1988).

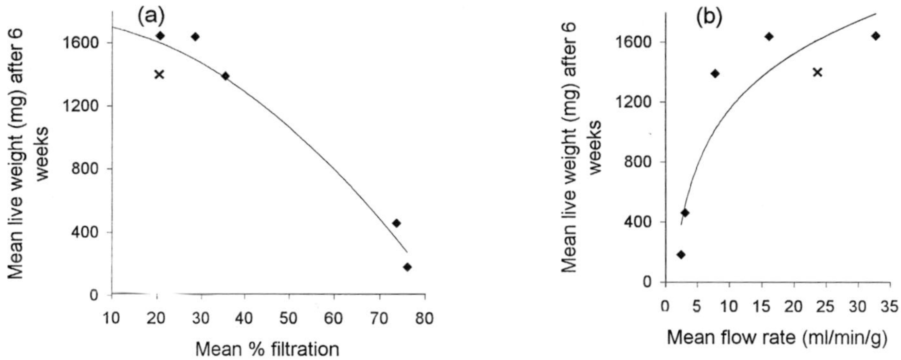

Fig. 2.10 Growth of 33 mg *Crassostrea gigas* in outdoor upwelling systems supplied with fertilised water in relation to (a) filtration of particles flowing through each upweller (the five levels of % filtration were calculated and applied daily from equation in Fig. 2.8 that predicted the required flow rate at the prevailing temperature, particle concentration and live spat weight), (b) the mean flow rates arising from the daily calculations. Values at X are data for oysters kept at a standard flow of 25 ml/min/g. Trend lines are shown (from Spencer, 1988).

rate and particle concentration for 10 mg *C. gigas* (Fig. 2.8). The same information, used to calculate the number of particles filtered (Fig. 2.9), shows that the interrelationship between particle concentration and flow rate leads to an optimum uptake of food, the level of which varies according to the interplay between these two factors. In these experiments, optimum uptake of food occurred at *c.* 20% filtration. Thus by adjusting flow rate to keep filtration to around this level, food uptake and growth (Fig. 2.10) can be optimised.

Eventually, there is total dependence on natural foods when juveniles are transferred to the sea in containers or on ground plots. Once there, the cultivator can do little to enhance feeding except by manipulating stocking densities and applying cleaning schedules to reduce siltation and fouling (see Chapter 4 on husbandry) to ensure that their stock receive an unhindered flow of food-bearing water. The consequence of filtering activity in terms of biodeposits may be little in high energy zones, where faeces and pseudofaeces are swept quickly away, but critical in sheltered areas, where they may accumulate on the beds and smother stock unless dealt with by the cultivator. These aspects are dealt with in Chapter 4.

References and bibliography

Brown, M. R. & Jeffrey, S. W. (1992) The nutritional properties of microalgae used in mariculture: an overview. In: *Proceedings of Aquaculture Nutrition Workshop, Salamander Bay, 5–17*

April, 1991 (eds G. L. Allan & W. Dall), pp. 174–9. NSW Fisheries, Brackish Water Fish Culture Research Station, Salamander Bay, Australia.

Espinosa, J. A. M. (1986) *The growth and survival of hatchery-reared Manila clam* (Tapes semidecussatus *Reeve*) and *American hard shell clam* (Mercenaria mercenaria) *in outdoor nursery pumped upwelling systems supplied with fertilized sea water.* M. Sc. Thesis, Plymouth Polytechnic, England.

Helm, M. M., Holland D. L. & Stephenson, R. R. (1973) The effect of supplementary feeding of a hatchery broodstock of *Ostrea edulis* L. on larval vigour. *Journal of the Marine Biological Association of the United Kingdom*, **53**, 673–84.

Holland, D. L. & Spencer, B. E. (1973) Biochemical changes in fed and starved oysters, *Ostrea edulis* L., during larval development, metamorphosis and early spat growth. *Journal of the Marine Biological Association of the United Kingdom*, **53**, 287–98.

Jørgensen, C. B. (1990) *Bivalve Filter Feeding: Hydrodynamics, Bioenergetics, Physiology and Ecology.* Olsen and Olsen, Fredensborg, Denmark.

Kennedy, V. S., Newell, R. I. E. & Eble, A. F. (eds) (1996) *The Eastern Oyster*, Crassostrea virginica. Maryland Sea Grant College, USA.

Malouf, R. E. & Castagna, M. (1989) Comparative biology of clams: Environmental tolerences, feeding and growth. In: *Clam Mariculture in North America* (eds J. J. Manzi & M. Castagna), pp. 23–73. Developments in Aquaculture and Fisheries Science, **19**. Elsevier, Amsterdam.

Millican P. F. & Helm, M. M. (1994) Effects of nutrition on larvae production in the European flat oyster, *Ostrea edulis. Aquaculture*, **123**, 83–94.

Morton, B. (1992) The evolution and success of the heteromyarian form in the Mytiloida. In: *The Mussel*, Mytilus: *Ecology, Physiology, Genetics and Culture* (ed. E. Gosling), pp. 21–52. Developments in Aquaculture and Fisheries Science, **25**. Elsevier, Amsterdam.

Newell, R. I. E. & Langdon, C. J. (1996) Mechanisms and physiology of larval and adult feeding. In: *The Eastern Oyster*, Crassostrea virginica (eds V. S. Kennedy, R. I. E. Newell & A. F. Eble) 185–229. Maryland Sea Grant College, USA.

Spencer, B. E. (1988) Growth and filtration of juvenile oysters in experimental outdoor pumped upwelling systems. *Aquaculture*, **75**, 139–58.

Utting, S. D. (1993) Procedures for the maintenance and hatchery-conditioning of bivalve broodstocks. *World Aquaculture*, **24** (3), 78–82.

Vahl, O. (1973) Efficiency of particle retention in *Chlamys islandica. Astarte*, **6**, 21–5.

Walne, P. R. (1974) *Culture of Bivalve Molluscs – 50 years experience at Conwy*. Fishing News Books, Farnham, Surrey, England.

Ward, J. E., Newell, R. I. E., Thompson, R. J. & McDonald, B. A. (1994) *In vivo* studies of suspension-feeding processes in the eastern oyster, *Crassostrea virginica* (Gmelin). *Biological Bulletin*, **186**, 221–40.

Winter, J. E. (1978) A review of the knowledge of suspension-feeding in lamellibranch bivalves, with special reference to artificial aquaculture systems. *Aquaculture*, **13**, 1–33.

Yonge, C. M. & Thompson, T. E. (1976) *Living Marine Molluscs*. Collins, London.

3 Hatchery culture of bivalve larvae and juveniles

Introduction

Although the artificial propagation of bivalves has been practised for over a millennium, hatchery cultivation, in which the environment required for repeated success is carefully controlled, is only a recent development. Modern hatchery methods, developed in the UK since 1960, have enabled relatively few hatchery units to produce routinely large numbers of juvenile bivalves. Currently there are two major commercial hatcheries in the UK, which in total produce around 100 million oyster seed per year, for sale at home and overseas.

Although hatcheries are essential commodities in climates where native species produce irregular spatfalls and where non-native species do not recruit at all, they are also commercially successful in countries, such as France, USA and Australia, where fisheries thrive on natural spatfalls.

This chapter is largely a synthesis of information published previously in laboratory leaflets (Utting & Spencer, 1991; Millican, 1997) prepared as guides to the procedures used in the hatchery rearing of bivalve molluscs in the United Kingdom. Much of the information on larvae culture and some of the figures used in the first leaflet were provided by our one-time colleague, Mike Helm. These procedures were developed at the Fisheries Laboratory, Conwy, during a research programme led by Peter Walne, which began in earnest in the late 1950s to early 1960s. Hatchery production of scallops is not yet commercially routine, but with the technology now available success is only a matter of time.

The species which are reared in commercial hatcheries in the UK and northern Europe include native and exotic (introduced or non-indigenous) species (Table 3.1).

The introduction of non-native species to the UK, with the exception of the American hard shell clam, were made via the quarantine facilities at the Fisheries Laboratory, Conwy. The broodstock were held in tanks of flowing sea water, under strict quarantine, and destroyed after spawning. The juveniles were reared in isolation with frequent histological checks by the Fish Disease Laboratory, Weymouth, to ensure their freedom from non-

Table 3.1 Bivalve species reared commercially in UK hatcheries

Species		Status	Year of introduction	Origin
European flat oyster	*Ostrea edulis*	native	–	–
Scallop	*Pecten maximus*	native	–	–
Carpet shell	*Tapes decussatus*	native	–	–
Pacific oyster	*Crassostrea gigas*	Introduced by MAFF, Conwy	1965, 1972 and 1978	Canada
Manila clam	*Tapes philippinarum*	Introduced by MAFF, Conwy	1980	USA
American hard shell clam	*Mercenaria mercenaria*	introduced (accidentally?)	1910–1920?	N America

indigenous parasites and disease causing organisms. Small numbers of disease-free juveniles were eventually supplied to British hatcheries for breeding purposes. The American hard shell clam, established in some of the south coast harbours, in Southampton Water and the Solent, is believed to have been introduced accidentally into England, possibly as discards from shipping arriving from America. These areas have provided broodstock for hatchery cultivation.

The main aspects of hatchery culture, shown in Fig. 3.1, comprising the husbandry of broodstock, larvae and juveniles within the hatchery, are described in this chapter. Broodstock conditioning, larvae culture and spat culture should be carried out in separate rooms to prevent mixing of sea water, and to reduce the risk of spreading disease or exacerbating problems associated with bad hygiene.

Conditioning of broodstock

For many species of bivalves in temperate coastal waters, sexual development starts in the sea when water temperature exceeds 10–14°C. At these temperatures, gametes develop during April–June and mature during July–August, to be retained until spawning is stimulated by increased temperatures or by thermal or handling shocks. With scallops, however, hatchery broodstock conditioning is best achieved during October–April, when the animals are developing gonad naturally. In northern European waters, where sea temperatures are rarely high enough to stimulate natural spawning of non-native species, mature gametes are retained into the winter and are then resorbed.

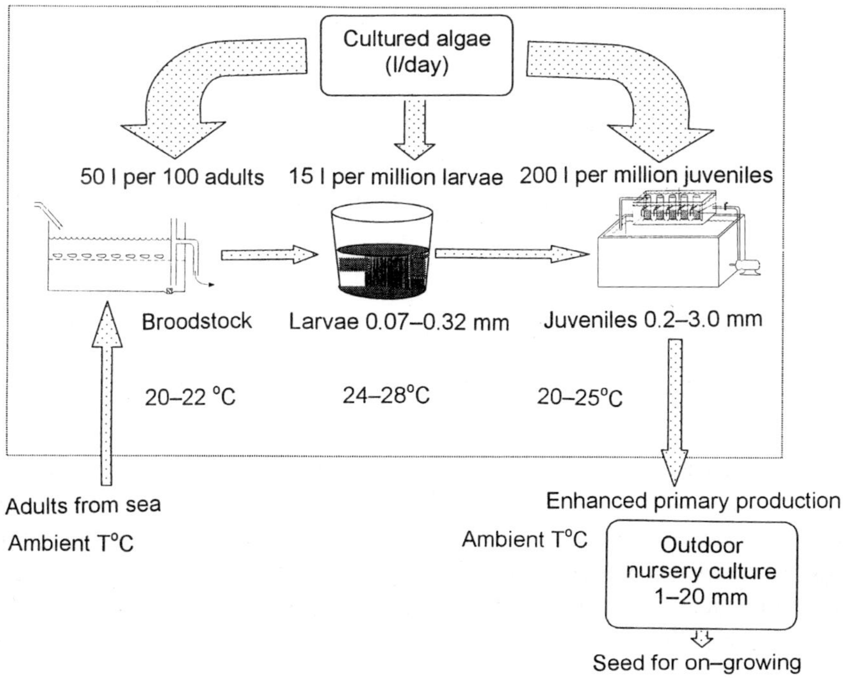

Fig. 3.1 Outline of hatchery culture.

Ripening of the gametes can be accelerated in the hatchery by keeping bivalves at elevated temperatures with a suitable ration of food. It is possible to produce mature adults in the winter and early spring before adults in the sea start their sexual development. Thus, bivalve species in spawning condition can be made available for most of the year.

Clams

Small clams, larger than 8 mm shell length (80 mg live weight), can be used as breeding stock, but since egg production increases with increase in size it is more usual to use large adults of 35–40 mm shell length (10–20 g live weight). These will spawn 5–8 million eggs on average, depending on their condition and the time of year when they are brought into breeding condition. All of the clams mature first as males (protandric hermaphroditism), and then with age either change sex to female or remain as males. Populations of 2- and 3-year-olds show a 50:50 sex ratio.

Oysters

Pacific oysters, like clams, mature first as males and then change sex to female or remain as males. Flat oysters also mature first as males but thereafter change sex at regular intervals. On rare occasions individual flat oysters remain either as males or females. At Conwy, mature adults of 30 g live weight (65 mm shell length for flat oysters, 70 mm for Pacific oysters) are used as broodstock. Females of these sizes may be expected to release 1 million larvae (flat oysters) or approximately 50 million eggs (Pacific oysters) per individual.

Scallops

In the UK *Pecten maximus* is sufficiently valuable to be considered suitable for hatchery production. Mature scallops (90–120 mm shell length) are used as hatchery broodstock. Since natural reproduction occurs in the spring and autumn, hatchery ripening of adults can be achieved by stocking them a month or so prior to these periods. This enables the breeding season to be extended for a larger part of the year. This species of scallop has both male and female gametes within the same individual, confined to an area known as the roe. The sperm develop in the proximal part (nearest to the attachment to the body) and the eggs in the distal (furthest from the body attachment) part of the roe. With ripening, the two zones become visually separable by strong colour differences of the red-orange ovary and the white testis. With careful management and the correct diet, broodstock may be expected to produce *c.* 20 million eggs per scallop.

Conditioning technique

Adult bivalves taken from the sea are brought into the hatchery and placed in a tray supported off the bottom in a tank similar to that shown in Fig. 3.2. Clams and scallops behave more naturally if kept in a substrate, such as coarse sand or gravel, in which they can bury themselves.

The sea water supply need not be filtered since the diversity and abundance of natural food species, which may be present in the unfiltered sea water, can be beneficial to the conditioning of the broodstock. Oysters and clams require a salinity of >25 practical salinity units (psu) and temperature of 22 ± 2°C. Scallops require a salinity of >30 psu and a temperature, initially of 9°C rising by 1°C each week to a maximum of 16°C. Water flow rate through the tank should be *c.* 25 ml/min per adult. This equates to a

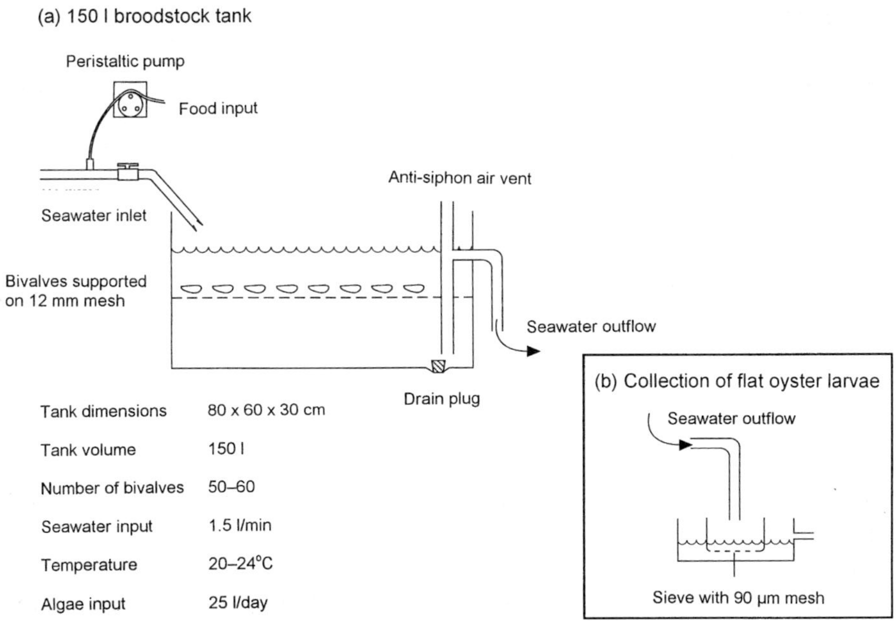

Fig. 3.2 The broodstock conditioning system.

stocking density of 8–12 g live weight per l of tank water capacity (i.e. 40–60 oysters or clams × 20–30 g live weight and 10 scallops × 150 g live weight in a tank of the dimensions shown in Fig. 3.2). The water must run to waste and not be recirculated or used again.

A commercial hatchery located in a disease-free zone (currently free of the diseases caused by *Bonamia* or *Marteilia*) would need a licence to deposit bivalves from non-approved zones or from overseas but would not wish to jeopardise its status by stocking broodstock from zones containing these diseases.

Cultured species of marine algae are used as the principal food supply during conditioning. *Tetraselmis suecica*, *Thalassiosira pseudonana* and *Skeletonema costatum* are useful species that can be cultured on a large scale. Each adult requires about 200 million cells of *Tetraselmis*, 2000 million cells of *Thalassiosira* or 1000 million cells of *Skeletonema* per day. A mixture of these species, on a proportional basis, is better than a single-species diet. A food ration, equivalent to 6% (oysters and clams) and 3% (scallops) of the initial dry bivalve meat weight of dry algae per day is recommended (see page 71).

Broodstock take 6–8 weeks (oysters or clams) and 5–11 weeks (scallops) to reach spawning condition in winter and early spring and a progressively shorter period with the encroachment of the natural breeding season.

The time taken to reach spawning condition can be calculated in terms of 'day-degrees' (D°):

$$D^\circ = d(t - t_o)$$

where D° = the number of corrected day degrees (i.e. above the threshold temperature),
d = the number of days to spawning,
t = the mean ambient temperature, and
t_o = the threshold temperature for the start of gonadal development.

In temperate climates where t_o for many bivalves lies between 10°C and 14°C, D° is between 300 and 500.

Spawning

Clams

Mature clams can be induced to spawn by applying certain stimuli. Unlike other bivalves reared in hatcheries, viable larvae cannot be obtained from artificially stripped gametes since the eggs need to undergo a maturation process during their passage down the oviducts. Various stimuli can be applied to induce spawning; the most successful are those that are natural and minimise stress to the animals.

Thermal cycling is a commonly used technique. Mature clams, taken from broodstock conditioning tanks, are cleaned externally and placed in a spawning tank. A shallow fibre glass trough, of approximately 150 × 50 × 15 cm deep, is suitable. The tank is fitted with a stand-pipe drain and two filtered sea water supplies, one heated to 18–20°C and the other to 28–30°C. The base of the trough is covered with black plastic sheeting to provide a dark background against which gametes being liberated can be seen quite readily (Fig. 3.3).

The trough is part-filled with the cooler water to a depth of about 10 cm and a small amount of cultured algae added to stimulate the clams to extend their siphons and start pumping activity. After 15–30 min the water is drained and replaced with water at 28–30°C, again with a small addition of algae. This water is drained after a similar time period and replaced with cooler water and the procedure is repeated. The number of cycles that are necessary to induce spawning depends on the readiness of the clams to spawn. In summer, clams may spawn within 1 hour but, earlier in the season, it may take 4–6 hours of cycling before the first animal spawns. Generally, if the clams do not respond within a 6-hour period they are returned to

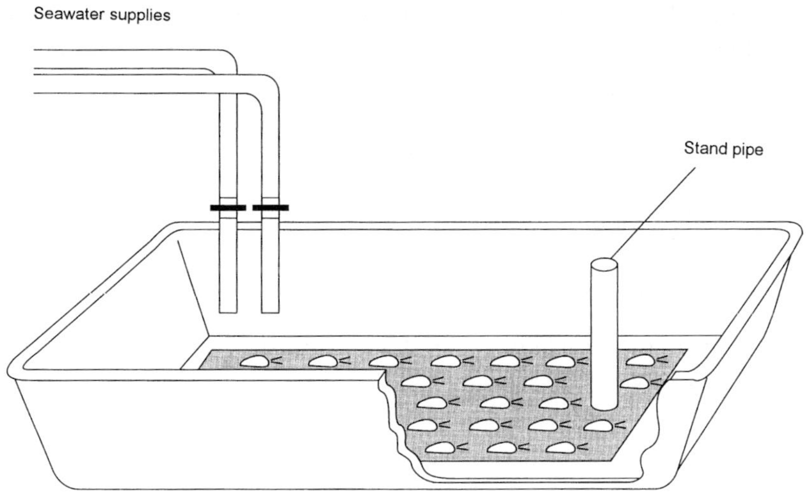

Fig. 3.3 The spawning trough.

the conditioning tanks for a further week. Clams may start spawning on the cool or warm part of the cycle, with males usually spawning first. Additional stimulus can be provided by the addition of stripped eggs or sperm. In Manila clams, the gonad is located at the base of the foot and can be seen when the mantle and gill tissues are lifted. If the gonad is carefully punctured with a Pasteur pipette and suction applied quantities of gametes can be withdrawn, which can then be placed in a small volume of filtered sea water. This suspension of gametes can then be dispensed towards the inhalant siphons (the one furthest away from the shell hinge line and which has the larger diameter aperture) of actively pumping clams with the pipette. The gametes, drawn into the mantle cavities by the pumping action of the clams, provide additional stimulus. When applied during the second warm cycle, there is almost always a spawning response within 0–4 hours of adding the suspension. When spawning occurs, eggs and sperm are expelled through the exhalant siphons.

When the first clams to spawn are males, it is good practice to remove them from the trough and leave them out of water until sufficient eggs have been collected from spawning females. This procedure is necessary since sperm ages rapidly with the consequence of 1-hour old sperm producing low fertilisation rates.

As each female begins to spawn, it is necessary to transfer it from the spawning trough to an individual spawning dish or beaker of about 1 l capacity containing about 400 ml of filtered sea water at 24–26°C (Fig. 3.4). The same procedure applies to spawning males, which can be identified as such

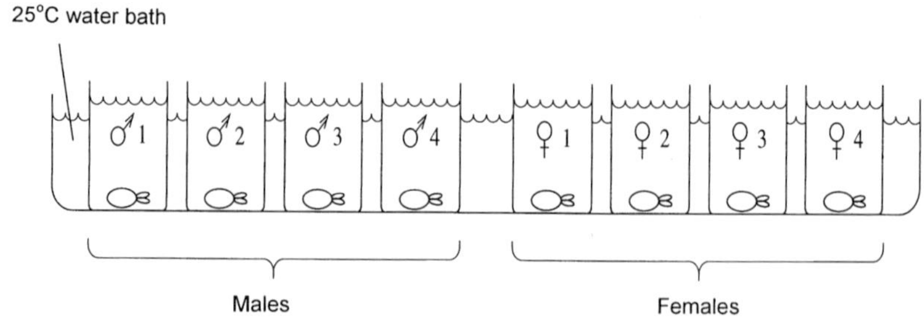

Fig. 3.4 Collection of clam eggs and sperm.

by the milky appearance of the continuous stream of sperm liberated from the exhalant siphon, as against the granular clumps of eggs shed from the female.

The time for completion of spawning for an individual clam is variable but gamete liberation rarely lasts for more than 15 minutes. It may, however, be necessary to remove a spawning female from its container and place it in a fresh one if large numbers of eggs have been released. The presence of dense concentrations of eggs in the water inhibits pumping activity and hence the further expulsion of eggs. In addition, the female may start to filter eggs out of suspension.

Eggs quickly settle to the bottom of the dish. When spawning has finished, any clumps of eggs can be separated by carefully pouring the contents of the dish through a 60 μm nylon mesh sieve, retaining the separated eggs on a 15 or 20 μm mesh sieve. The eggs are then gently washed into a clean glass or plastic container with filtered sea water at 25°C. When first spawned the eggs are pear-shaped but rapidly assume a spherical shape once in contact with sea water. Eggs from the different females are collected separately to provide the opportunity to assess quality visually, using a microscope at about ×100 magnification. Batches of eggs which do not round off after about 10 minutes in sea water, or which are not of a universally dense, granular appearance, should be discarded. Reproductive development in the clams is not completely synchronous so that, at any one time, eggs spawned by different females will be at different stages of maturation.

When separation and examination of the eggs are complete, batches of eggs that appear in good condition can be pooled. Sperm from separate males are similarly pooled. It is good practice to use eggs from at least six females and sperm from a similar number of males for a production run. This ensures a good genetic mix in the offspring, depending on the degree of heterozygosity of the parents.

It is desirable to estimate the egg numbers before fertilisation, since development will be impaired if the density of embryos per unit of volume beyond early cleavage exceeds certain limits. This density and the method used to determine egg numbers are described later (see page 56).

Pacific oyster

Thermal cycling can also be used to induce Pacific oysters to spawn. A faster method, however, provided unlimited broodstock are available, is to remove gametes artificially from the adults (Helm & Millican, 1977). If the flat shell valve is removed, the gonad can be punctured with a Pasteur pipette and the gametes withdrawn. Eggs or sperm are dispensed into 400 ml of filtered sea water at 24–26°C in a 1 l beaker, using a separate pipette for each oyster. Eggs are poured carefully through a 60 µm mesh sieve and collected on a 15 or 20 µm mesh sieve. Treatment of eggs and sperm is then the same as that for clams.

Native flat oyster

This species is larviparous (i.e. egg fertilisation and early development of larvae take place within the gill cavity of the parent. The larvae are approximately 170 µm in shell length when shed into the surrounding water. Once liberated into the broodstock tank, the larvae swim upward towards the water surface, where they appear as a grey film. This avoids rapid loss of larvae from the tank since the overflow pipe drains from near the bottom (Fig. 3.2). Larvae may be collected from the water surface by scooping the upper layers of the tank water with a plastic jug and pouring the contents onto a 90 µm mesh sieve. As a safeguard against loss of larvae from overnight releases, a 90 µm mesh sieve may be left in position permanently (Fig. 3.2). The sieve should be kept clean and checked regularly during the day for the presence of larvae. Treatment and estimation of numbers of larvae are described later (see p 56).

Scallops

Thermal cycling may also be used to induce scallops to spawn. The procedure is similar to that for clams. Water temperature, however, should cycle between 16°C and 20°C. Usually sperm are released first. Once spawning starts, the scallop should be removed to its own container of water at

16°C. When sperm release stops, the scallop should again be isolated. This prevents the possibility of self-fertilisation should the scallop subsequently release eggs.

Fertilisation

Clams, Pacific oysters and scallops

After pooling the eggs and sperm in separate containers, fertilisation is carried out by adding 2 ml of a dense sperm suspension to each litre of the egg suspension. The sperm should be less than 30 minutes old from the time of spawning and the eggs less than 60 minutes old. Sperm should not be used to fertilise the eggs from the same scallop. After addition of the sperm, the contents of the container should be gently agitated and then allowed to settle for 60–90 minutes. Within this period (25°C for clams and oysters; 16°C for scallops), the fertilised eggs divide, first into two cells of equal size and then into four cells composed of one large cell capped by three much smaller cells (Fig. 3.5). Assessments of the percentage of eggs developing normally can be made using a relatively low-power microscope (×20 magnification). Fertilisation rates invariably exceed 90%.

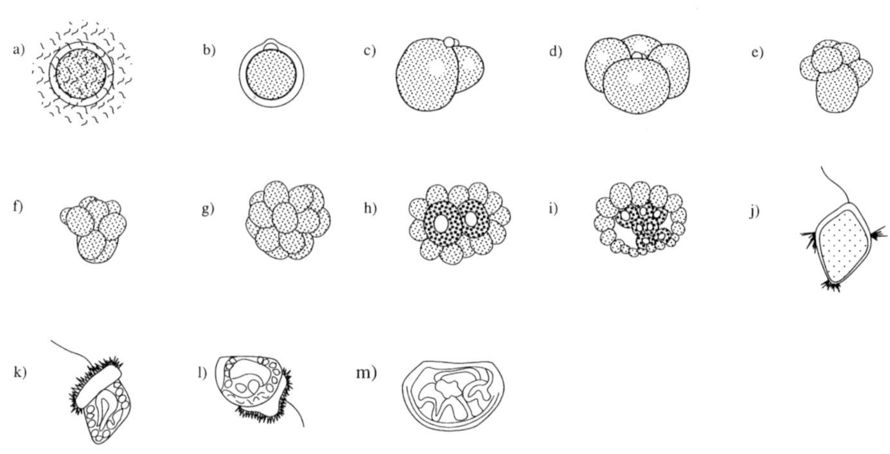

Fig. 3.5 Development of the Manila clam embryo.

Embryonic development

Clams, Pacific oysters and scallops

Fertilised eggs are left to develop to the fully-shelled, D-veliger stage in flat-bottomed rearing vessels (Fig. 3.6). This early veliger stage is known as the D-larva stage because of the characteristic 'capital D' shape of the shell.

Vessels of 100–250 l are ideal, preferably made from pigment-free, translucent polyethylene. New vessels must be filled with sea water and allowed to soak, with weekly changes of water, for 3 months before use. This allows toxic substances which may be harmful to larvae to leach from the plastic. Once ready to use, the vessels are filled with sea water, filtered by a rapid sand filter, a diatomite filter or similar device, which will retain particles greater than 2 µm in diameter. The water is heated (25°C for clams and oysters; 16°C for scallops) and salinity adjusted to 25 psu for Pacific oysters, 26–30 psu for clams and >30 psu for scallops. It is often beneficial to prepare the water 24 hours in advance, with the addition of 1 mg of EDTA (disodium salt as used in the preparation of algae culture medium) per litre

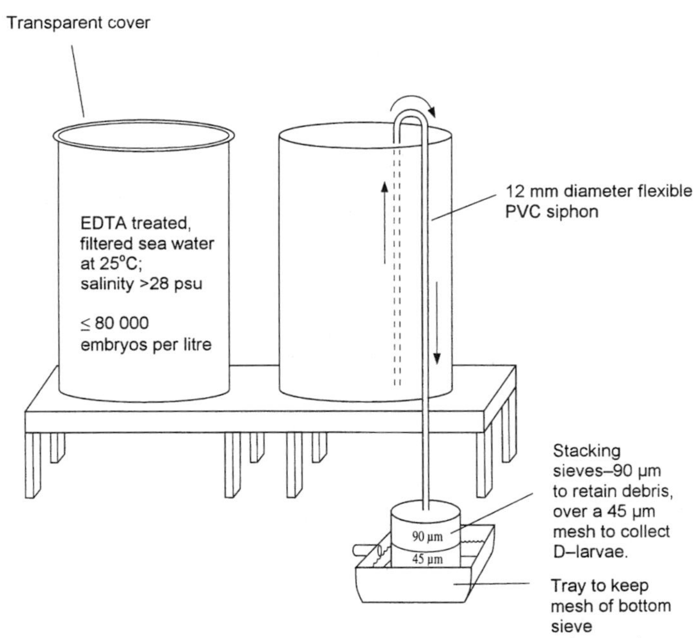

Fig. 3.6 Rearing vessels for embryos and larvae: the method for recovering D-larvae is shown.

(Utting & Helm, 1985) and with vigorous aeration. The air flow is turned off during embryonic development.

Embryos are stocked in the vessels about 2 hours after fertilisation at a maximum density of 50 000 per litre. Fully developed D-larvae are recovered, by the method illustrated in Fig. 3.6, 24–72 hours later.

In cultures of up to 200 l capacity, 30–50% of the initial number of embryos may be recovered as perfectly formed D-larvae. Imperfectly formed D-larvae (i.e. those with misshapen shells) are unlikely to develop further. The normal D-larvae should have a mean shell length of about 90–5 μm (clams), 70–5 μm (Pacific oysters) and 100–15 μm for scallops (measured by monocular microscope, ×100 magnification fitted with an eye-piece graticule calibrated against a micrometer slide).

Normal D-larvae are retained by a 45 μm mesh sieve and the number recovered is estimated as described below.

Estimating numbers of eggs and larvae

Care is needed when handling eggs and larvae. During transfer to other containers it is important to ensure that the mesh of the sieve is always submerged (Fig. 3.7a). Much of the equipment used in the hatchery needs to be specially made. For example, sieves are prepared from PVC tubes or high-impact, rigid polystyrene horticultural box containers. The bases are removed from the latter and nylon monofilament meshes affixed with PVC solvent cement. Perforated plunger agitators and counting slides can be made in the workshop from acrylic or PVC sheet and rod. Specific equipment can be purchased from suppliers of laboratory equipment, including adjustable volume automatic pipettes (0.1–1.0 ml and 1.0–5.0 ml ranges are useful) and various plastic measuring cylinders from 25 ml to 2000 ml in volume.

A useful procedure to follow is:

- wash and sieve the eggs or larvae and transfer them to a 1 or 2 l measuring cylinder,
- make the volume up to the graduation mark with filtered sea water,
- set an automatic pipette at 0.5 ml and, while gently agitating the contents with a perforated plunger, take three replicate samples from the cylinder, making sure that the eggs or larvae are evenly distributed in the water column during sampling (see Fig. 3.7b and c),
- dispense the samples into the compartments of a counting chamber (Fig. 3.7d),
- count the eggs or larvae using a binocular microscope,

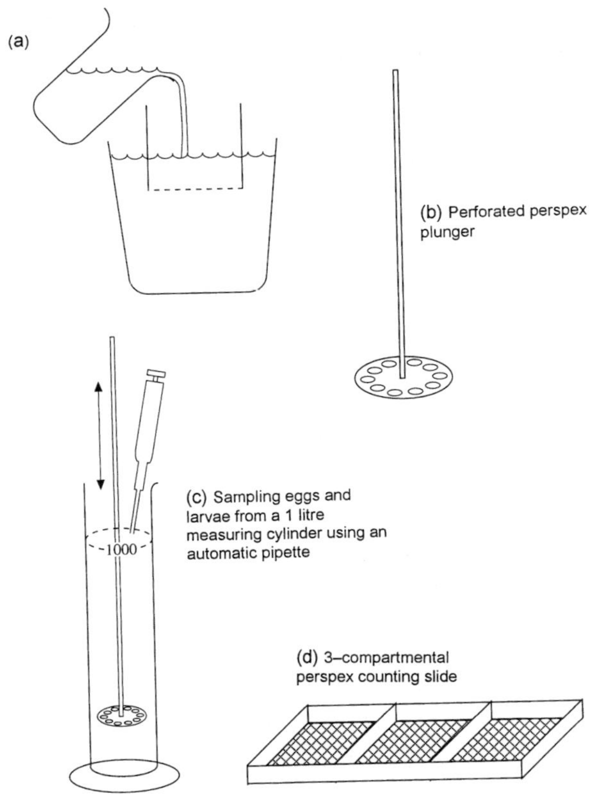

Fig. 3.7 Equipment used for estimating egg and larvae numbers.

- calculate the total numbers of eggs or larvae according to the following example:

  counts of eggs or larvae in samples = 236, 254, 239 (mean = 243)
  volume of sample = 0.5 ml
  volume of water in cylinder = 1000 ml

  therefore, the total number of eggs or larvae $= \dfrac{243 \times 1000}{0.5} = 486\,000$

With very high numbers, smaller volume samples or a larger volume measuring cylinder, or a combination of the two, may be used. In the case of eggs, which are very delicate, it may be easier to transfer the egg suspension to a larger container, e.g., a 10 l plastic bucket. In this case, make up the volume to the calibration line and withdraw samples while gently agitating the contents of the bucket with a larger diameter, perforated plastic plunger.

An electronic particle counter, such as one of the models supplied by Coulter Counter, though expensive, is an invaluable tool in determining numbers of eggs and larvae and is also of value in determining densities of unicellular algae in culture.

Rearing of larvae

Basic conditions

The larvae can be grown equally well in flat-bottomed vessels (Fig. 3.6) or in conical-shaped, glass-fibre tanks fitted with bottom drains (Fig. 3.8). D-larvae can be grown up to 15 000–20 000 per litre but growth and survival are improved considerably at densities below 10 000 per litre.

Growth is improved by ensuring that the water is sufficiently turbulent to keep the larvae in suspension. However, too little or too much aeration can depress the growth of larvae. Aeration by a single central aerator, at flow rates of 200 l per hour for 75 l vessels (depth of water = 45 cm) and 350 l

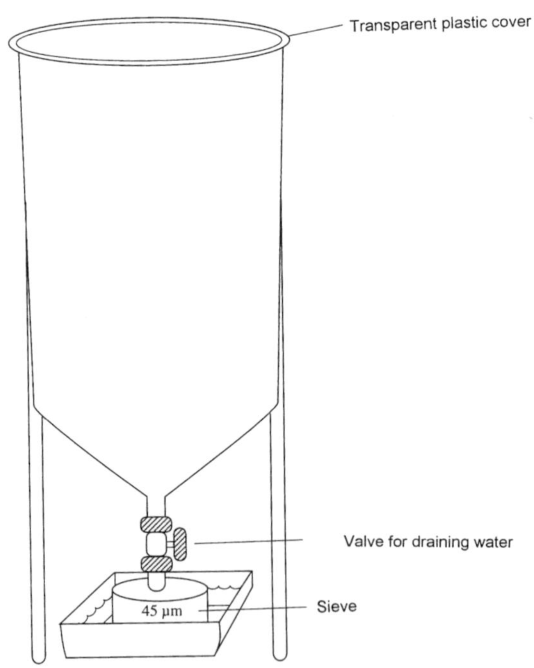

Fig. 3.8 Conical rearing vessel for larvae.

per hour for 350-litre vessels (water depth = 76 cm), is suitable for near optimal growth (Helm & Spencer, 1972). Compressors that produce carbon or oil in their air supplies are unsuitable for the hatchery. The air is filtered to 0.45 µm particle size by a series of cartridge filters of decreasing porosity. This is needed to reduce airborne contaminants that may contain harmful micro-organisms.

Sea water for the larvae is filtered to about 2 µm particle size and is heated to 24–26°C for oyster and clam larvae and 16°C for scallop larvae. Salinity should be 25 psu for Pacific oysters, 26–30 psu for clam larvae and above 30 psu for flat oyster and scallop larvae. Following filtration, sterilisation of the water by ultraviolet light to reduce bacterial contaminants is optional but advisable.

Clam, Pacific oyster and scallop D-larvae, and newly-released flat oyster larvae are at the stage where they need feeding with live unicellular algae. Suitable food species include *Chaetoceros calcitrans, Thalassiosira pseudonana* (3H), *Isochrysis galbana* (or 'Tahiti Iso'), and *Tetraselmis suecica* (for larvae > 120 µm in length). Methods for the mass production of these algae are described elsewhere (Laing, 1991). Diets of mixed algae are beneficial. A suitable diet for the D-larvae is a mixture of *Chaetoceros* and *Isochrysis*. Food rations are usually quoted as the number of algae cells per microlitre (cells per µl). When a mixture of *Chaetoceros* and *Isochrysis* is fed, the most suitable densities are 125 cells per µl (50 cells per µl for scallops) and 50 cells per µl respectively. *Isochrysis* and *Chaetoceros* can be fed alone at 100 cells per µl and 250 cells per µl respectively. With a two-species diet, each alga species is added at 50% of the cell density at which it is fed as a single species diet. Larvae greater than 120 µm mean shell length grow best on a three-species mixture composed of the following proportions:

33 cells per µl *Isochrysis* + 83 cells per µl *Chaetoceros* + 3.3 cells per µl *Tetraselmis*

Volumes of the alga species needed to achieve the required cell densities are calculated from the following formula:

$$\text{Volume (ml)} = \frac{\text{Required cell density (cells per µl)} \times V \times 1000}{\text{Cell density of cultured algae (cells per µl)}}$$

where V is the volume of the rearing vessel in litres.

Over-feeding is equally damaging to the performance of larvae as under-feeding. With high densities of larvae, it is necessary to add the total daily ration in two or more feeding sessions to avoid grossly exceeding the optimum cell density.

The number of algae cells eaten per day, by oyster and clam larvae of different shell lengths grown at 24°C, is shown in Fig. 3.9. One cell of

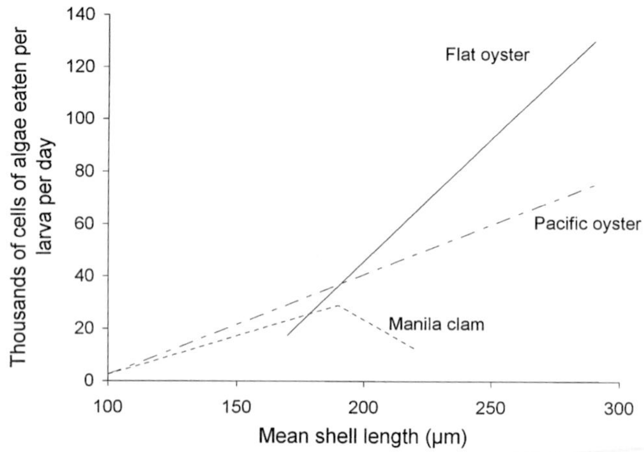

Fig. 3.9 Feeding of larvae: number of algae cells (equivalent in size to *Isochrysis*) eaten per day at 24°C.

Tetraselmis is calculated as being equivalent in cell volume to 10 cells of *Isochrysis* or 25 cells of *Chaetoceros*, and values are shown in *Isochrysis* cell equivalents. Numbers of cells eaten per larva per day increase as the larvae grow (Fig. 3.9), but with clams they decrease once a size of 190 μm is reached (see p 62).

The procedure for starting a larval culture is as follows:

- Fill a clean larvae rearing vessel with filtered sea water at about 25°C for clams and oysters or 16°C for scallops, and at the appropriate salinity.
- Abnormally high mortalities of larvae may be caused by bacteria. Thus, it is advisable to treat the water with UV light before filling the vessels. If mortalities persist, a broad-spectrum antibiotic may be used under veterinary prescription.
- Add clam, scallop or Pacific oyster D-larvae at a density of 10 000 larvae per l and add newly liberated flat oyster larvae at a density of 2000 larvae per l.
- Calculate the volumes of cultured algae needed (see p 59) and add to the vessels to provide the required food ration.
- Aerate the water gently for D-larvae but more vigorously for larger larvae (>140 μm). Flow rates of air for optimum growth and spat production vary with size and shape of the container. As a guide, an aeration rate of 200 l per hour and 350 l per hour are required for conical containers with 75 l and 350 l capacities (Helm & Spencer, 1972).

- The culture is now left for 24 hours before further husbandry is required.

Husbandry of larval cultures

Since larval cultures are operated as static water systems (i.e. without a continuous exchange of water), they require daily attention. The daily addition of algae and the accumulation of waste products during the larvae's normal feeding, respiration and excretion can lead to harmful water quality in the vessel. At initial stocking densities of below 10 000 larvae per l, it is usual to change the water in the culture vessels three times per week. At higher stocking densities, more frequent water changes are necessary to maintain satisfactory water quality.

Husbandry at water changes

The procedure is similar to that for embyos (see p 55):

- Water is siphoned from the vessel through a sieve with a mesh sufficiently large to allow larvae to pass through but which retains debris. Initially, a 200 μm sieve is ideal. The larvae are retained on the 45 μm mesh of the lower sieve.
- Wash any remaining larvae from the vessel into the sieve.
- Wash the vessel with a sponge and hot detergent solution and rinse well.
- Refill the vessel with appropriately treated sea water.
- Grade the larvae by washing them through a stack of sieves of descending apertures with filtered sea water. Suitable mesh sizes are shown in Fig. 3.10.
- Observe the appearance and activity of small samples of retained larvae from each sieve with a microscope. Discard any fraction containing mainly dead larvae (i.e. empty shells and those containing decomposing tissues) which may be retained on the lower sieves. Healthy larvae have yellow-brown coloration with a dark digestive gland.
- Wash the healthy larvae into a measuring cylinder. Sample (see p 56) to determine the total number surviving. The addition of a few drops of formalin to the samples will immobilise the larvae.
- Return the larvae from the cylinder to the rearing vessel and adjust the aeration to the required level.
- Feed.
- Repeat this procedure at the next water change.

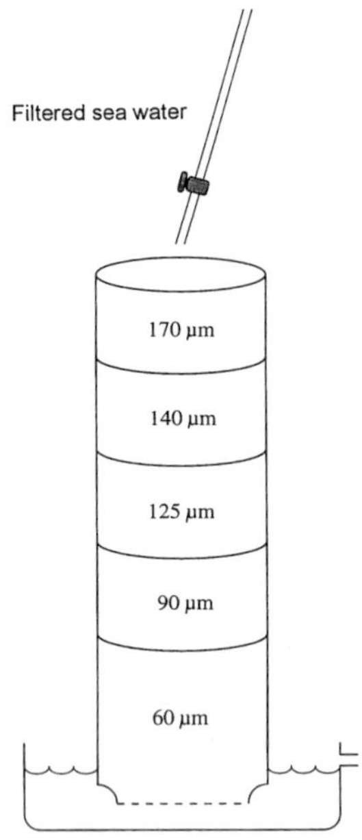

Fig. 3.10 Stack of sieves suitable for Manila clams. Oysters near to settlement require meshes of 210 µm and 245 µm and Pacific oyster D-larvae require meshes of 35 µm and 45 µm in addition to those shown.

Husbandry between water changes

This consists of replacing the algae grazed down in the previous 24 hours. A sample of water is taken from the vessel and the residual algae cells are counted, either microscopically using a haemocytometer, or with a Coulter Counter.

Daily records are kept of culture temperatures and residual food cell concentrations (Fig. 3.11). Volumes of algae required to replace that grazed, are calculated using the formula given previously.

Growth of larvae

Information on the growth of larvae is shown in Fig. 3.12.

Clams

Growth in the first 8 days, when the larvae are actively swimming, is rapid. Growth rates decrease as larvae reach the pediveliger stage (*c.* 200 µm) when they gradually change to a more sedentary habit, crawl by means of a foot, and undergo metamorphosis (*c.* 230 µm) to the juvenile or spat stage. Young

Date:	Day:	Vessel volume (l):	Expt. no.	T°C 26.1
	12	125	9	salinity (psu): 30

Mean shell length (μm) = 211.4	% eyed =		
Size class (μm)	Frequency	no.	diff.
160			–31
170	I	1	–4
180	III	3	–9
190	IIII	4	–8
200	~~IIII~~ ~~IIII~~	10	–10
210	~~IIII~~ ~~IIII~~ ~~IIII~~ ~~IIII~~ IIII	24	×
220	IIII	4	+4
230	III	3	+6
240	I	1	+3
250			+13
	mean x = 215 – 3.6 = 211.4 μm	n = 50	= 13 –31 × (100/50) = –3.6

Water treatment	
Answer	✓ or ×
water change?	✓
water filtered?	✓
UV treated?	✓
EDTA added?	✓
Antibiotics?	×
Other? (specify)	

Larvae observations	
colour:	very good
Activity:	very good
sub-sample vol:	0.5 ml/2 l
counts:	183,177,180
total number:	720000

Feeding					
Residual algae		algae added			
Species	cells/μl	Species	cells fed	ml added	cells/μl
ISO	15.6	T.ISO	100	954	13100
CHAET		CHAET	100	236	53000
TET	1.2	TET	10	521	2400

Oyster spat numbers	
sub-sample vol.	ml/ l
counts:	
total number:	
grand total:	

Larvae grading		35	45	61	90	124	140	170	210	236	265	μm sieve
approx. %					5 (poor)	10	45	40	few			
Retain	✓				×	✓	✓	✓	✓			
Discard	×											

Notes: large numbers of pediveligers

Fig. 3.11 Specimen of daily record sheet.

spat tend to crawl up vertical surfaces and cluster together using their byssal threads.

Oysters

Pacific oyster larvae grow rapidly to metamorphosis during their first 12 days in culture. Flat oyster larvae, however, are quite large (*c.* 170 μm) when

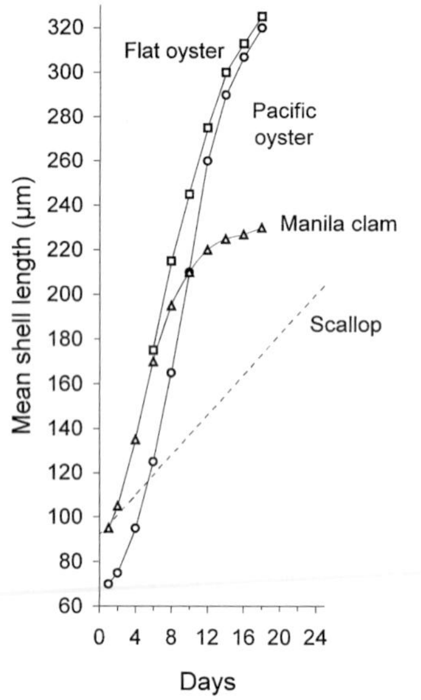

Day 0 : fertilisation
Days 1–3: D-larvae
Flat oyster larvae released from parent at *c.* 170 µm

Fig. 3.12 Growth of larvae: oysters and clams at 25°C and scallops at 16.5°C.

liberated from the parent, and hence their free-swimming stage is reduced to about seven days at 25°C. Before metamorphosis at 230–300 µm, oyster larvae cement themselves to a suitable substrate. This phase is called 'settlement'. The settlement substrate used in the hatchery at Conwy is a matt-surfaced, black PVC disc placed flat on the bottom of the rearing vessel. Some hatcheries overseas use scallop shell ground into fragments of approximately 180 µm in diameter. These are added to the rearing vessel prior to settlement and act as settlement surfaces for individual larvae. The chemicals epinephrine and nor-epinephrine, added at concentrations of 10^{-4}–10^{-5} M, can be used to settle and metamorphose without the provision of a settlement surface (Coon *et al.*, 1986).

Larvae that are ready to settle can be recognised under the microscope by the presence of a dark, pigmented eyespot on the tissues inside the shell. When approximately 70–80% of the larvae have reached the 'eyed' stage, the PVC disc is added to the rearing vessel and the vessel is illuminated. Larvae swim away from the light and attach to the disc. 'Settled' oysters are removed daily from the disc by scraping carefully with a razor blade and the numbers are estimated (see p 56).

Scallops

Scallop larvae grow relatively slowly owing to the low temperature of culture, taking 22–25 days to reach metamorphosis at *c.* 250 µm shell length. Metamorphosis is a critical period during which disturbance and handling should be kept to a minimum. Once 50% of the larvae have developed an eyespot and a foot, they are transferred to a mesh tray (50 × 60 × 17 cm deep) with a 150-µm nylon mesh at a stocking density of 20 000 larvae per tray.

Survival of the larvae

During the period of cultivation, which may last for 2–3 weeks depending on species, about 50% of the larvae may be lost owing to natural mortality and by the intentional removal of the slow growers during the sieving process. A further 50% may be lost during metamorphosis, leaving about 25% of the initial number of larvae stocked. These values, however, are highly variable. Optimum survival occurs early in the year, with the poorest survival usually in the summer, again associated with higher temperatures and bacterial infection.

Husbandry of pediveligers

Clams

Clam pediveligers are usually left in their rearing vessels until day 16 and, thereafter, those retained on a 170-µm sieve are transferred to re-circulating sea water systems of 1000–2000 l capacities operated by air-lift (Fig. 3.13). Smaller larvae are retained in the larvae vessel, with routine husbandry, until they are large enough to transfer.

An air-lift downwelling recirculation system, suitable for holding clam pediveligers during metamorphosis, is shown in Fig. 3.13. The feeding and husbandry for this type of system is similar to that for the larvae vessels (see p 61). As pediveligers complete metamorphosis and start to grow rapidly, they require additional food above the basic ration. A 50% supplement may be added initially and this can be increased further with evidence of higher grazing rates.

By day 35, many of the spat should be retained on a 440-µm mesh. These spat may be transferred to an upwelling system (Fig. 3.14). It is sufficient to grade pediveligers and early spat once each week. More frequent grading or

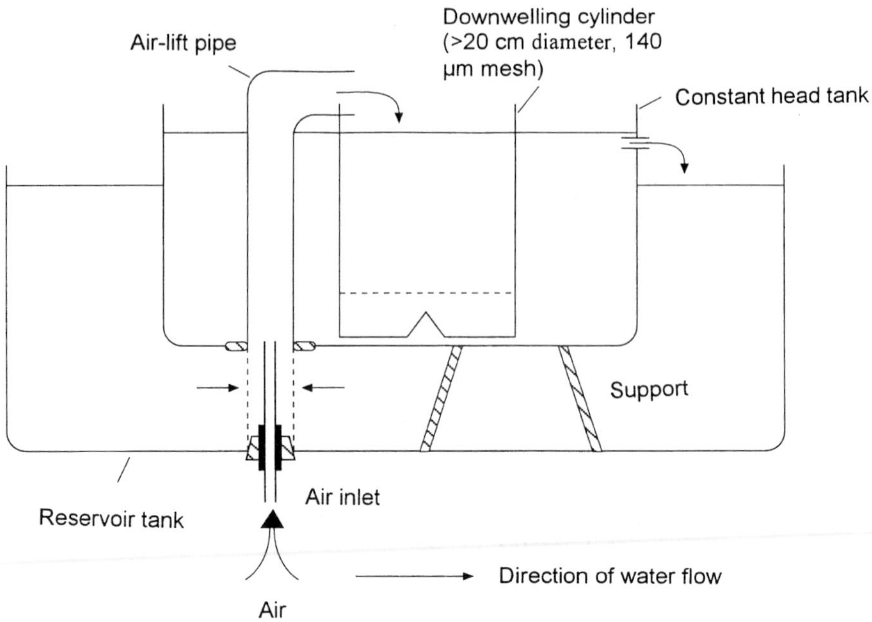

Fig. 3.13 An air-lift downwelling, recirculation system suitable for clam pediveligers during metamorphosis.

excessively rough handling may suppress growth. Meshes of 210 µm, 265 µm, 350 µm, 440 µm, 660 µm, 850 µm, 1000 µm and 1500 µm are useful for grading. The tanks need washing with detergent and hot water at each water change. Periodic cleansing of the valves and pipework by internal brushing is necessary to remove accumulated algae debris. The spat may be rinsed with sufficient pressure to remove faeces and uneaten food. A 100-l downwelling system is suitable for growing 0.5 million small spat to a size which will be retained on a 440 µm mesh. Spat are particularly vulnerable during this stage, and although 70% survival may be achieved in good conditions survival is likely to vary widely.

Scallops

Scallop pediveligers (*c.* 170 µm shell length) are placed in mesh trays (see p 65) stored in shallow glass fibre tanks supplied with recirculated filtered sea water at 16°C. After passing through the holding tanks, the water is collected in reservoir tanks at the bottom of the system from which it is pumped back to the scallops. A typical system with a capacity of 1600 l is

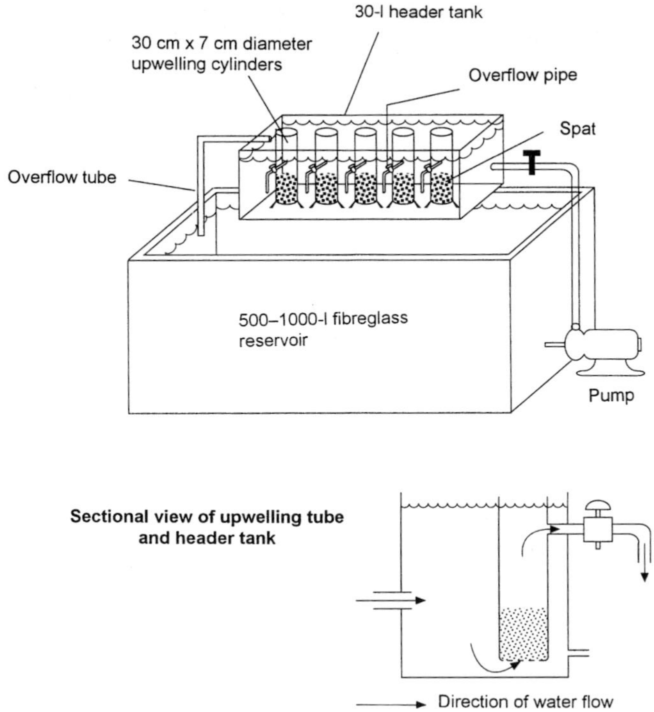

Fig. 3.14 A nursery pumped upwelling, recirculation system for holding spat.

shown in Fig. 3.15. The scallops are fed with 5–10 l of a mixture of at least two species of algae. The water is changed three times per week. Two water changes are made without disturbing the scallops, but on the third change the baskets are removed and gently hosed to clean the meshes and to wash away accumulated debris.

Husbandry of spat

Nursery upwelling systems for oysters and clams

Upwelling systems are suitable for oysters immediately after settlement and for clams after completing metamorphosis in the downwelling system (see p 65). A 210-μm mesh, fitted to the base of the upwelling cylinder, is appropriate for newly-settled oyster spat. Once clam spat are retained on 440-μm mesh, they are fairly robust and easy to handle. They are no longer able to swim but are extremely mobile and can crawl up vertical surfaces. As a

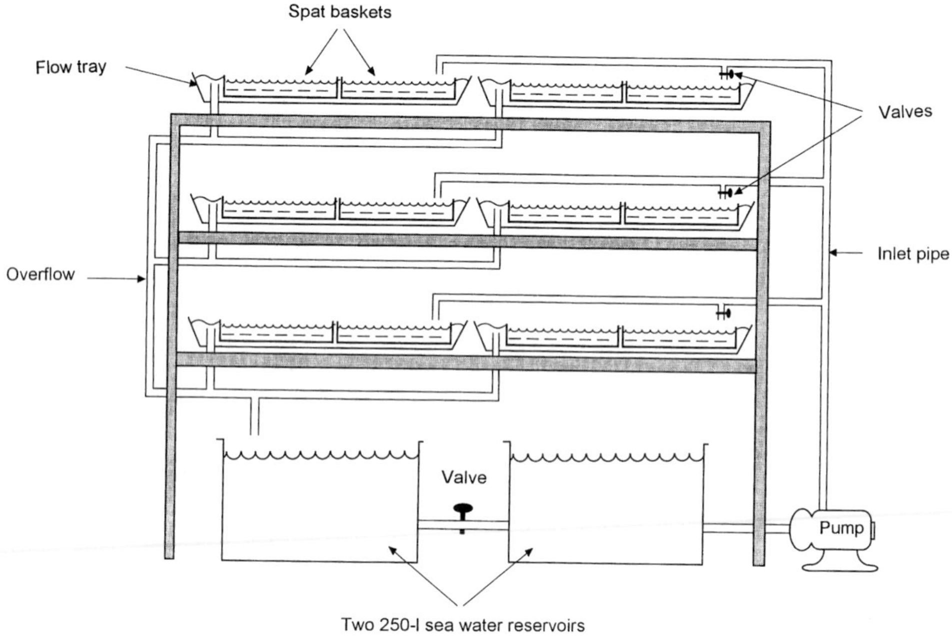

Fig. 3.15 A recirculating system (1600-l capacity) used for rearing scallop spat.

consequence many will climb above the water/air interface, where they become stranded and desiccate and eventually die. Care must be taken to dislodge these stray spat every day.

Two types of upwelling system are shown (Figs 3.14 and 3.16) in which water flow through the cylinders is induced by creating a difference in the head of water levels. The water flow may be induced by an electrically-driven pump (Fig. 3.14), or with an air-lift (Fig. 3.16). Valves may be fitted to the overflows of the upwelling cylinders, since spat growth is strongly affected by water flow rate. Flow rates of 20–50 ml/min/g live weight of spat should be used.

Sea water, coarsely filtered through a 45-μm sieve, may be used so that the spat can benefit from the presence of naturally occurring algae in addition to the added cultured algae. The water is kept at 22–25°C and is changed three times each week. The systems are thoroughly cleaned at each water change and the spat washed (see p 65). Since individual spat within a batch show widely differing growth rates, grading is usually carried out once each week where the objective is to group spat of a similar size together in an upwelling cylinder.

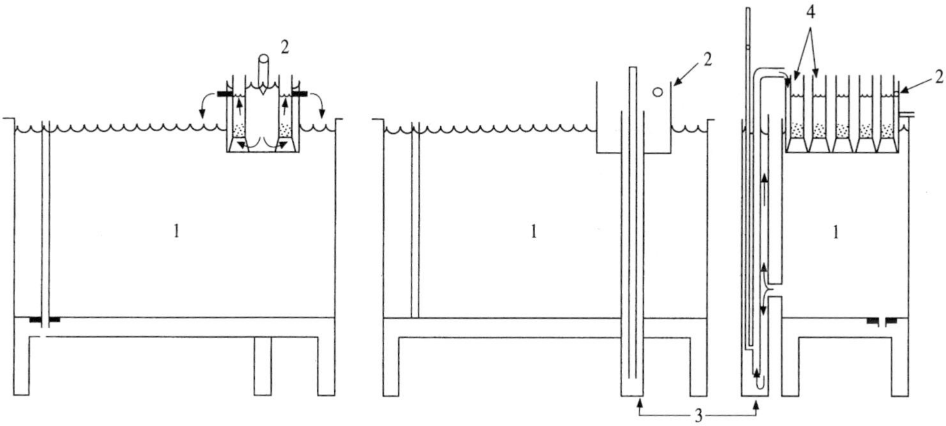

Fig. 3.16 Air-lift upwelling spat system. (1) 3000-l reservoir tank; (2) spat tank; (3) air-lift apparatus; (4) upwelling tubes.

Nursery upwelling systems need to be of relatively large volume, since only small biomasses of spat can be grown successfully per unit volume of water. The consequences of over-stocking are shown in Fig. 3.17. It is good practice to regulate the biomass to a maximum of 200 mg/l of the system volume. For example a 2000 l system can be used to grow a maximum of 400 g live weight of spat, irrespective of the mean live weight of the spat. Four hundred grams live weight can be made up of 400 000 × 1 mg (live weight) spat of 1.5 mm mean shell length or 40 000 × 10 mg spat of 4.4 mm mean shell length. To help with calculating the numbers of spat that can be held at different mean live weights, a length to weight conversion graph is given (Fig. 3.18).

Examples of stocking, flow rates and food requirements for systems having tank volumes similar to those in Figs 3.14 and 3.16, are given in Table 3.2.

A nursery system for scallops

After spending 2 weeks in the pediveliger system, most of the larvae have metamorphosed into spat, which attach by their byssus threads to the walls and bottom of the container. The spat are transferred to a system essentially the same in design and handling as that used for the pediveligers but with a procedural difference in water management. Water at 16°C, coarsely strained through a 45-μm sieve to exclude larger debris yet retain naturally occurring phytoplankton, is added continuously at a rate of 60 l per hour.

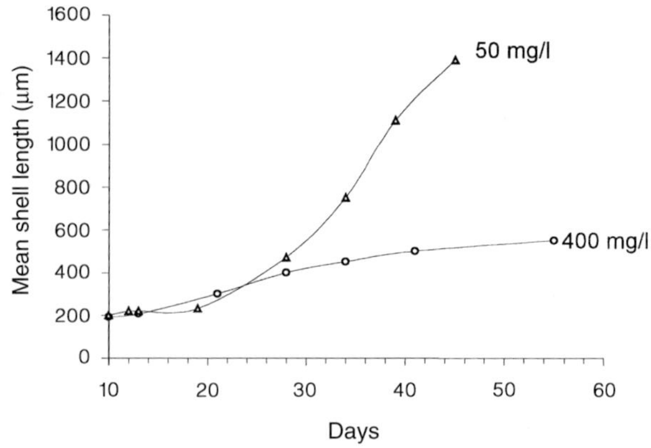

Fig. 3.17 Growth of Manila clam spat at initial stocking densities of 400 mg/l and 50 mg/l.

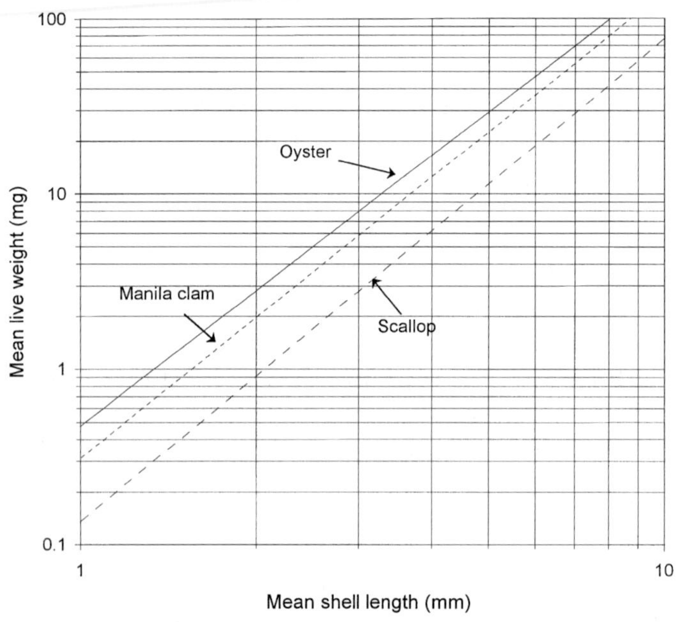

Fig. 3.18 Length/weight relationship of Manila clam, scallop, and Pacific and flat oyster spat.

Surplus water is run to waste. Each day, 10–20 l of a mixture of algae is added. At this time, the water supply is stopped for 6–8 hours to avoid wasteful feeding.

Once the seed reach 5 mm, after *c.* 6 weeks in the system, they can be transferred to larger mesh trays in tanks supplied only with continuously

Table 3.2 Spat stocking, water flow rates and food requirements for spat upwelling systems of different volumes

Tank volume (litres)	Optimum spat capacity at 200 mg/l (g)	Minimum flow at 30 ml/min/g (l/min)	Food at 2×10^6 cells of *Tetraselmis*★ mg/week (l/day)
500	100	3	28.5
1000	200	6	57
3000	600	18	171

★ *Tetraselmis* equivalents at 1 million cells/ml, fed as three portions per day

flowing sea water. After a further 4 weeks, about half of the spat grow to 10–15 mm, a size suitable for transfer to pearl nets in the sea providing the temperature is above 12°C.

Food requirements of oysters and clams

Spat growth is largely influenced by the amount of food available for feeding. The ration is calculated as dry weight of algae (mg) fed over a week, per mg of live weight of spat at the beginning of the week. The amount of growth which may be expected with Pacific oysters fed different rations of *Tetraselmis* is shown in Fig. 3.19. One million cells are equivalent to 0.2 mg dry weight (2.5 million cells to 0.5 mg etc.).

Since feeding is based on the weight of spat in the system, their biomass must be determined each week. Feeding a ration of 0.4 mg dried algae per mg (live weight) of spat per week provides good spat growth without excessive cleaning problems associated with higher levels of feeding. The amount of food required daily is calculated from the following formula:

$$F = (S \times 0.4)$$

where F is the dry weight of algae required per day (mg) and S is the live weight of spat (mg) at the beginning of each week.

For example, 200 g of spat (200 000 mg) require 200 000 × 0.4/7 = 11 429 mg (11.43 g) dry weight of algae per day. To determine the number (in millions) of algae cells equivalent to this weight of food, divide by the weight (in mg) of one million cells. Table 3.3 gives the weights of cells of commonly used species of algae.

Therefore, to feed 11.43 g of *Skeletonema* daily, 11 429/0.032 = 357 156 million cells would be required. To convert this number of cells into the volume of algae culture (in litres) required, divide by the concentration of cells in the culture (cell per μl). Thus, if *Skeletonema* culture was harvested

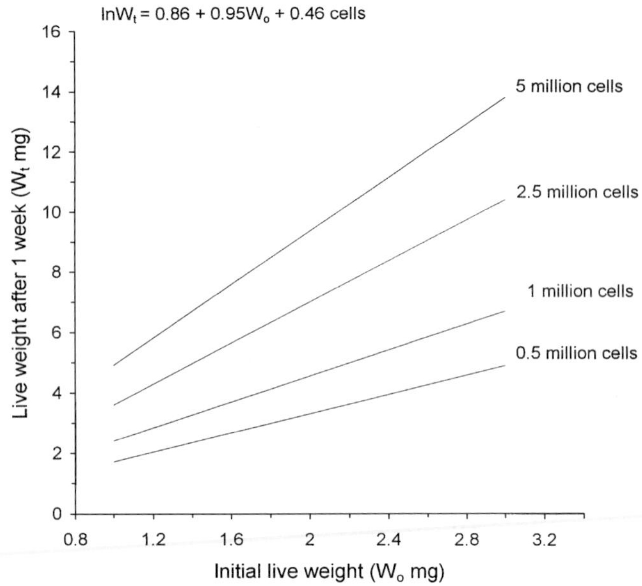

Fig. 3.19 Growth of Pacific oyster spat fed different rations of *Tetraselmis* (millions of cells per week per mg of spat). Spat were kept in trays supplied with water entering as splash jets from above. Temperature was 23–25°C and salinity 25 psu.

Table 3.3 Dry weight of common species of algae

Species of alga	Weight (mg per million cells)
T-Iso	0.02
Skeletonema costatum	0.032
Chaetoceros calcitrans	0.007
Chroomonas salina	0.13
3 H (*T. pseudonana*)	0.02
Tetraselmis suecica	0.20

at 7000 cells per µl, then a volume of 357 156/7000 = 51.02 litres of this culture would be needed for feeding each day. The daily food requirement (V, litres) of spat in upwelling systems can be calculated from one equation, thus:

$$V = \frac{(S \times 0.4)}{7 \times W \times C}$$

where S is the live weight of spat (mg) at the beginning of the week, C is the cell concentration (cells per µl) of the culture, and W is the weight (mg) of one million algae cells (Table 3.3).

It is preferable to feed a mixture of two or more species of algae, with one of the best combinations being a mixture of 50% *Skeletonema* and 50% *Tetraselmis*. To determine the amount of each species to feed the equation is used to calculate the total volume (V) of each type of alga which would be required if it would be the only food, and the result is divided by 2.

Live algae substitutes as food have been the source of investigation for many years. In the USA, live algae reduced to pastes by centrifugation are available commercially. These may be kept in cold storage for up to about a week and are reconstituted in sea water before use. In the UK spray-dried algae produced quasi-commercially support good growth of spat when provided as a 50% substitute with live algae. Although the technology for manufacturing good quality dried algae exists, spray-dried algae are not yet available commercially. Fully artificial diets incorporated into microcapsules sufficiently small to be ingested by larvae have been manufactured, but their constitution falls far short of that needed to match the growth and survival rates achieved with natural algae.

Growth and survival

Examples of growth of Manila clam and Pacific oyster spat are shown in Fig. 3.20. Pacific oysters grow faster than Manila clams but both species usually reach 5 mg (2–3 mm) within 6 weeks in indoor nurseries. Survival rates are variable but 50–90% may be expected providing spat quality at settlement is satisfactory and adequate feeding and husbandry are maintained.

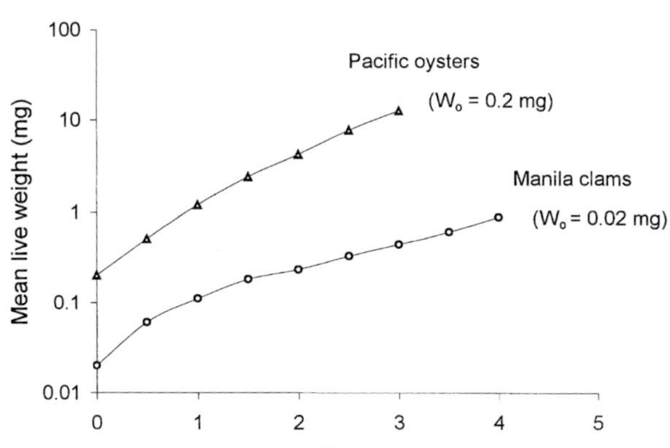

Fig. 3.20 Growth of Manila clam and Pacific oyster spat in the same upwelling system fed with *Tetraselmis* at a rate of 0.28 million cells per day per mg live weight of spat. Water temperature averaged 22°C, salinity 30 psu and flow rate 30 ml per min per g of spat per upweller.

It is usual for hatcheries to sell clam and oyster spat, or put them into outdoor nurseries, at a mean shell length of 2–3 mm (5 mg), because of the considerable expense of growing them further under closely controlled conditions indoors. If there is a big difference between the temperature of the hatchery rearing systems and the ambient temperature of the sea, subsequent spat survival will be greater if they are acclimated gradually before transfer by decreasing the temperature in the spat system by 2°C per day.

General hygiene within the hatchery

The initial selection of a suitable site on which to build the hatchery is of utmost importance to its subsequent success for production of bivalve seed. A supply of high quality sea water, free from pollution and organisms that cause disease, is essential. The design of the hatchery is critical, to ensure that aspects of hygiene and safety are of the highest standards.

Before a hatchery is built, thought should be given to its design to ensure that problems associated with fouling are minimised. The first area of concern lies within the piped sea water supply, which because it carries a biologically active medium, may over a period of time become fouled with marine invertebrates such as mussels and hydroids. These may reduce the rate of water flow through the pipes and in some circumstances affect water quality. This, in turn, may interfere with the performance of larvae culture in the hatchery. For these reasons, the length of pipe-work supplying the hatchery should be kept to a minimum, and should be fitted with easily uncoupled joints to facilitate cleaning. In circumstances where the sea water is supplied intermittently rather than continuously, e.g. for algae culture and larvae culture, the pipe-work should be designed to avoid dead areas where sea water may collect between use. It should also be made of steam-sterilisable material (glass or polypropylene) to enable sterilisation to be carried out immediately after use, on a daily basis if necessary. The fitment of draining points also ensures that the pipe-work can be kept free of culture medium in between usages.

Hygiene in the algae, larvae and spat culture rooms is of prime importance. The rooms should be kept free of unnecessary equipment that may clutter the working space and prevent thorough cleaning. The floors of these areas are invariably used for the disposal of water containing debris from the cultures, and should therefore be well drained and not allowed to harbour stagnant pools. It is good practice to hose the floors daily with fresh water and frequently sterilise them by dousing with a weak solution of hypochlorite. Frequent sterilisation of the equipment used for handling larvae and spat is also good practice.

Although routine attention to hygiene is likely to reduce the risks of bacterial diseases of larvae, there are occasions when mortalities within cultures and between cultures may persist over a period of time. At this point it is advisable to undertake a thorough cleansing and sterilisation of the hatchery equipment and environment to kill off the causative bacterial organisms. It may also be necessary to replace the broodstock after sterilising the tankroom since the origin of the problem may lie in this area.

Conclusions

Despite many years of research, problems in the reliable hatchery production of bivalves still occur. Successful hatchery production of larvae and spat is related more to the skill and experience of the staff than to the excellence of the facilities and equipment. A keen awareness of hygiene is essential. In general terms greater success can be expected in the cooler months of the years, when it is necessary to condition adults to spawn outside of their normal breeding season. Higher mortalities of larvae, and on rarer occasions of spat, can be anticipated in the summer and are associated with higher temperatures that produce greater bacterial infection.

This chapter provides a brief review of the techniques used for the reliable hatchery culture of bivalve species most commonly reared in the UK. The techniques are appropriate for the rearing of native flat oysters (*Ostrea edulis*), Pacific oysters (*Crassostrea gigas*), Manila clams (*Tapes philippinarum*), native palourdes (*T. decussatus*), scallops (*Pecten maximus*) and the American hard shell clam (*Mercenaria mercenaria*). With modifications, the techniques can also be used for the culture of other bivalve species. In other parts of the world, hatchery techniques basically conform to those described here. Some departures are noteworthy. In the USA, billions of eyed Pacific oyster larvae are sent by express mail to oyster farms remote from the hatchery. They are placed in settling tanks with mussel shell or other collecting surfaces, on which to metamorphose shortly after arrival. Although at some risk, the success rate of settlement of the larvae is sufficiently high to encourage the use of this method. Its advantages lie in the low cost of the seed and low transport costs.

References and bibliography

Coon, S. L., Bonar, D. B. & Weiner, R. M. (1986) Chemical production of cultchless oyster spat using epinephrine and nor-epinephrine. *Aquaculture*, **58**, 255–62.

Helm, M. M. & Millican, P. F. (1977) Experiments in the hatchery rearing of Pacific oyster larvae (*Crassostrea gigas* Thunberg). *Aquaculture*, **11**, 1–12.

Helm, M. M. & Spencer, B. E. (1972) The importance of the rate of aeration in hatchery cultures of the larvae of *Ostrea edulis* L. *Journal du Conseil international pour l'Exploration de la Mer*, **34**, 244–55.

Laing, I. (1991) *Cultivation of Marine Algae*. Laboratory Leaflet, MAFF Directorate of Fisheries Research, Lowestoft, **67**.

Millican, P. F. (1997) *The Hatchery Rearing of King Scallop* (Pecten maximus). Laboratory Leaflet, CEFAS, Lowestoft.

Utting, S. D. & Helm, M. M. (1985) Improvement of sea water quality by physical and chemical pre-treatment in a bivalve hatchery. *Aquaculture*, **44**, 133–44.

Utting, S. D. & Spencer, B. E. (1991) *The Hatchery Culture of Bivalve Mollusc Larvae and Juveniles.* Laboratory Leaflet, MAFF Directorate of Fisheries Research, Lowestoft, **68**.

4 Cultivation of hatchery-reared oysters in the sea

Cultivation: Procedures and principles

Since the early 1960s, the appearance world-wide of commercial bivalve hatcheries has required the development of new cultivation techniques to ensure that the large quantities of tiny spat produced are grown as quickly and safely as possible to market size. This chapter provides information on some aspects of the procedures and methods for the cultivation of hatchery-reared Pacific and European flat oysters that have been developed at the Fisheries Laboratory, Conwy.

Hatchery seed

Source

Oyster seed may be bought from commercial hatcheries and nurseries in which hatchery-reared stock has been grown to a larger size. The usual range of sizes offered for sale are 3–30 mm (0.005–3 g) at prices ranging from £7 to £30 per thousand depending on size (1999 UK price list).

Sizes

Commercial suppliers refer to oyster size by the length of the shell, and often sell seed graded to sit on a particular mesh size. Commonly used sizes of mesh in tray construction are 2 mm, 5 mm, 9 mm and 18 mm, with those dimensions referring to the length of one side of square mesh aperture. Other, intermediate, mesh sizes are available from some manufacturers.

To ensure that all the population is retained, the oysters should be sieved with a mesh 30–50% larger than that used on the tray. When measuring oysters, it is simpler to count and weigh a sample than to measure shell length. Thus, oyster size in this chapter is usually expressed as the average live weight in grams (g) or milligrams (mg). The relationship between shell length (longest axis) and live weight for Pacific oysters (Fig. 4.1) is similar to that for flat oysters.

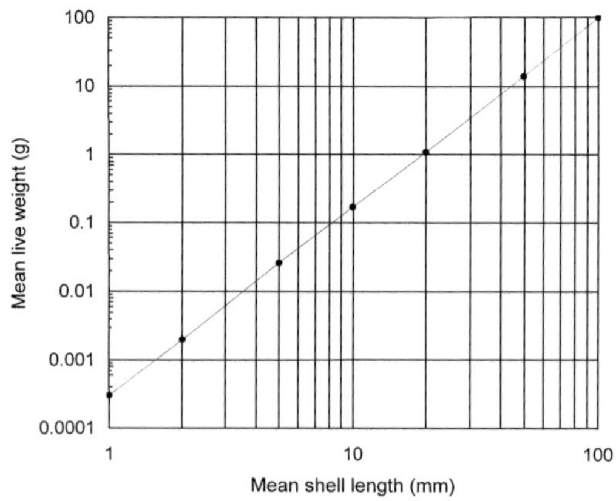

Fig. 4.1 The relationship between shell length and live weight of Pacific oysters.

Quality

Spat quality is judged by the capacity of the spat to survive and grow well. This depends largely on the early welfare of the spat in the hatchery and may vary considerably between batches, months or years. It is difficult to estimate visually or quantitatively, but on receipt the seed should be tightly closed and contain a low proportion of dead or empty shell. A reputable supplier may replace seed that shows high mortality shortly after re-laying, so long as he is satisfied that it has been treated carefully by the purchaser. A few hundred spat kept separately in a tray and counted regularly for a month or so is a useful ploy by the purchaser to enable him to monitor the quality of a batch in its early days after planting.

Trial plantings of Conwy-reared Pacific oyster seed, smaller than 8 mm, in trays in the Menai Strait, Wales, showed that survival after 6 months averaged 77% (3–4% mortality per month). The percentage of batches with different survival rates was as follows:

% of batches	Average % survival after 6 months
82	89
11	52
6	33
2	14

With careful handling, a high proportion of batches should show good survival. Survival of flat oysters in trays is usually lower than that of Pacific oysters. For example, 32 samples of Conwy-reared flat oysters planted in trays at seven sites, had an average monthly mortality of 8% (range 3%–15%), which is two to three times higher than that for Pacific oysters. This phenomenon may be linked to a tendency for oysters to form layers in trays with the smooth flatness of flat oysters, compared with the cupped shape and crinkly surface of Pacific oysters, preventing food and oxygen-laden water reaching those in the lower layers. The survival of flat oysters is very much improved if they are kept in forced upwelling systems in the first year of their life.

Although spat survival improves with increasing size, there is no ideal size for the purchase of seed. Much depends on the experience, equipment, strategy and effort of the cultivator. The higher initial cost of larger seed must be judged against the improved yield and eventual savings in labour and equipment costs. As a general guide, a cultivator may expect 70% survival of Pacific oysters in the first year, and thereafter 90% per year to market size using careful cultivation practice.

Planting season

Seed may be planted successfully in trays in any month from April to September. The cultivator should recognise that there is some risk in planting during October to March, with success depending largely on climatic conditions and site. At low tray densities, 0.5 g (15 mm) Pacific oysters bought in the spring should grow to 10–20 g by the end of their first season and 50 g in their second season. However, 0.01 g (3–4 mm) oysters planted in July should grow to 1–5 g in their first season and probably will not reach 50 g until their third season. At high stocking densities and with less frequent servicing schedules, growth to market size may be prolonged by a year or more.

The growth potential of Pacific oysters of various sizes varies with site (Fig. 4.2). For example, in 1982 one season's growth of 1 g oysters ranged from 4 g to 70 g depending on site. The two poorer sites are now known to have been affected by TBT contamination. The best site was a disused quarry that benefited from high summer temperatures and an abundance of natural food. Site differences were less pronounced with the flat oyster because of its slower growth and higher tolerance to TBT.

Systems and materials

Hatchery- and nursery-reared oysters require some form of protection until they reach about 10 g in size. Unprotected juveniles are vulnerable to predation, mainly by shore crabs, to strong tidal currents and wave action and to siltation. The effects of these hazards can be minimised by a variety of

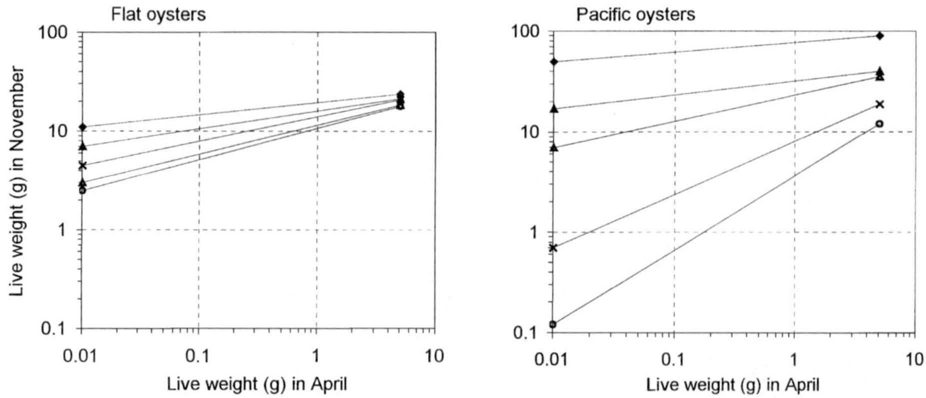

Fig. 4.2 Growth of flat oyster and Pacific oyster seed at five sites in England and Wales.

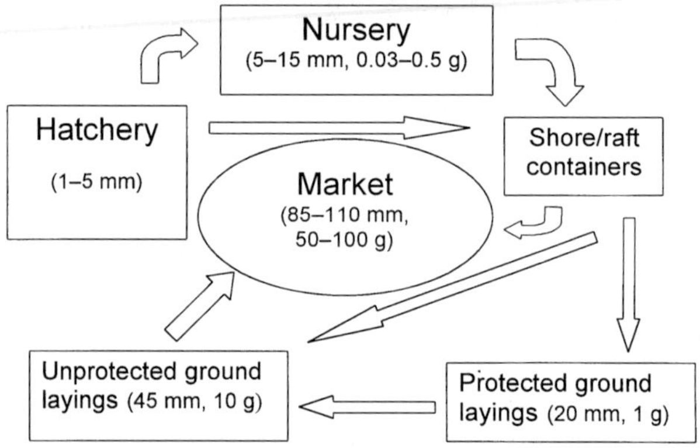

Fig. 4.3 Flow diagram showing some options for growing hatchery seed to market size.

methods of cultivation that enable the cultivator to safeguard his stock through the critical early stages. The flow diagram (Fig. 4.3) shows some of the options available for protecting oysters when growing them to market size.

Intertidal culture

Oysters may be grown intertidally on the shore in containers on trestles. They may also be grown on the ground within protective, crab-proof, fenced plots or on unprotected plots if the oysters are large enough to resist crab attack. For optimum growth, the oysters should be submerged all of the time, but clearly this is impracticable where access is required from the shore.

Fig. 4.4 Various types of trays attached to a trestle at LWMST.

Fortnightly visits are usually possible if the racks or ground lays are positioned at low water mark of spring tides, where they are submerged for about 90% of the time. This allows access for several hours during a spring tide. Oysters that are exposed for 3–4 hours during low water periods of neap tides are too high up the beach for growth but should store well during the summer months.

Tray or bag (pôche) culture is the most commonly used method for protecting oysters. Home made trays using a wooden frame covered with plastic netting are probably the cheapest available to the cultivator who does not cost his labour. These may be single- or multi-compartmental as preferred. Light gauge timber (50 mm × 25 mm) is suitable for tray construction, since it is not too buoyant in its early days in the sea, nor too heavy for handling after becoming waterlogged. A convenient size of tray for small spat (3–4 mm) has two compartments each 30 × 30 × 5 cm deep (Fig. 4.4). The dimensions of the compartment should not be much larger than this size, since oysters do not remain evenly spread but wash into heaps along one side or corner of the tray. The plastic mesh is usually attached permanently to the bottom with wooden or plastic battens, while that on the top is temporarily battened for regular access during servicing. A 1.5 mm mesh is suitable for enclosing 10 mg (3–4 mm) seed but once these reach about 100 mg (8–10 mm) they can be transferred to 6 mm mesh. Using a size of mesh appropriate to the size of the oyster is essential to ensure a good exchange

Fig. 4.5 Plastic bags on trestles showing temporary sealing rod at one end.

of water through the tray. If wooden-framed trays are to be stacked more than two high, 5 cm thick spacers should be put between the trays to allow a good flow of water to the oysters.

Untreated timber is quickly eroded away in most coastal sites in the UK by the boring action of the gribble worm (*Limnoria* spp.), which is a small crustacean. The useful life of timber in the sea can be increased by several years by protecting it with a proprietary brand of chemical preservative. However, chemically preserved timber must be avoided or used with caution at sites with poor water exchange to minimise risk of chemical contamination. Preservatives containing TBT must not be used. Commercially available trays take the form of 'lay-flat' plastic mesh flexible tubes or bags with openable ends (Fig. 4.5) or rigid open-topped plastic trays with moulded edges that enable them to be stacked (Fig. 4.6). The bags, approximately 1 m × 0.5 m × 6 cm deep, may require an insert to provide rigidity, but are also available with reinforced side meshes to prevent their collapse. The bags

Fig. 4.6 Stacks of NWP trays.

are available with various mesh sizes ranging from 4 to 18 mm. Stackable trays are also available in various sizes and meshes; dimensions up to 1 m × 0.5 m are a convenient size for handling. Five trays with oysters and a sixth tray empty as a lid constitute a suitable number for a stack with an overall height of about 30 cm. Larger numbers of trays within a stack make it unwieldy to handle and can cause reduced growth in the upper levels. Rigid trays are probably more suitable for growing small spat in their first year on the shore, whereas bags are more popular and widely used for growing larger, older oysters.

Metal racks (trestles) are probably the most durable and versatile means of supporting trays on the foreshore. They should be positioned so that the trays are held 0.3–0.5 m above the sea bed. Two bars are suitable for supporting rigid trays but three may be necessary for flexible bags. Stabilising a rack to prevent it from toppling in excessive wind or wave action can be achieved by pushing its legs into the soil, by weighting the legs with concrete or by interconnecting adjacent parallel racks to provide a wide base. Rubber bands, cut from old motor car inner tubes, provide a quick and secure method of fixing trays to racks. The band is looped around the bar of a rack, stretched over the tray and clipped by a plastic covered metal hook to its opposite partner, using four bands per tray. When securing stacks of trays, several bands may be joined together to cover the extra distance involved. Stacks of open-topped trays are best secured by strapping them together with one or two nylon straps. This also aids handling, especially when transporting trays up or down the beach. Because of their collapsible nature, bags should not be stacked more than one container deep. An ideal arrangement, adopted in some areas of France, is to secure each bag to the frame along one edge only and to slightly lap the other edge over its neighbour. At intervals of 1 or 2 months, the bags are flipped over 180° to redistribute the oysters, to dislodge silt, and to smother weed on the previously exposed upper surface.

Ground culture on subtidal or low intertidal lays is the traditional way of cultivating wild-caught flat oysters in the British Isles. However, hatchery-reared seed require some form of protection during their early years in the sea to ensure good survival. Oysters smaller than 10 g live weight (*c.* 45 mm) are vulnerable to crab attack and require protection for 1–2 years until this size is reached. Although tray cultivation is the most widely practised and probably the most successful way of protecting oysters during this vulnerable period, other methods have been tried. Covering ground plots of oysters 0.5–5 g in size with plastic netting may be suitable for short periods, but netting causes rapid deposition of silt which may smother the oysters. Also, as found with clam lays covered with thin, flexible netting, crabs are able to crack open the shells unless two layers of netting are used.

Protective fencing, used traditionally in France and the USA, can protect oysters and clams against attack by crabs (see Chapter 10). However, they can be erected only intertidally, in sheltered waters without a strong tidal flow. The walls of the fence are made from 10 mm plastic netting, about 50 cm high, with their bottoms buried in the soil to a depth of about 15 cm. The top of the fence has a crab-proof overhang, made from smooth plastic material that faces outwards at 90° or 45° to the vertical position. Experimental fences, up to 40 m × 20 m in size erected on muddy ground in the Menai Strait, were effective crab barriers.

Although highly successful for growing mussel seed, they were less so for Pacific oysters smaller than 10 g owing to unacceptably high mortalities caused by smothering from silt. Pacific oysters larger than 10 g are virtually crab-proof and, hence, can be transferred to unprotected ground layings. Satisfactory large-scale intertidal pens can be made on soft ground by enclosing plots 2 m wide with 12 mm plastic mesh walls, 15 cm high, buried 7.5 cm into the soil and staked at intervals. These walls confine the stock to the plots. The number of stock determines the length of the plots. Walkways 0.5 m wide may be left between plots for access during planting, servicing or harvesting (Fig. 4.7). Similar plots can be prepared on hard ground, with suitable barriers to prevent washing out of the stock. Once oysters have been re-laid on the ground, they require little attention until market size is reached.

Fig. 4.7 Ground plots of Pacific oysters.

The preparation of the ground depends largely on the site and requirements and resources of the cultivator. Ground can be cleared or improved for re-laying in various ways. For example, depositing shell or other suitable material may harden soft, muddy soil but the effort may be prohibitively costly for large areas. Soft, muddy ground can be used for Pacific oyster culture without further treatment, but it may be of benefit to harden narrow roadways for foot or vehicular access to key points on the beach from whence a hand-pulled sledge provides a means for carrying equipment and stock.

Floating culture

Rafts

These may be convenient for cultivating oysters in suspended trays in some sheltered waters as an alternative to shore-based trays. Oysters on rafts are continuously submerged and access to them is independent of tidal low water. The water must be sufficiently deep to prevent the trays from grounding and must have a good tide run of 1–2 knots (50–100 cm per second) to ensure a good flow of water through the trays. Raft design can be relatively simple with two flotation compartments joined by timber spars from which the trays are suspended. Small rafts, capable of supporting a stack of trays, can be linked together as modular units, but larger rafts, supporting many stacks, may be considered to be more appropriate. Raft trays are supported in frames that must be sufficiently strong to carry heavy loads and withstand the stresses and strains imposed by swift tides and wave action.

Trials in the Menai Strait showed that stacks that swing with the prevailing current offer no growth advantage over those held rigidly in the water column. Frames must be shackled to a raft with galvanised chain and lifting gear may be required to haul them. Comparative trials in the Menai Strait have shown no consistent difference in oyster growth in trays on a raft or on the shore. Other sites may prove to be consistently more favourable for one or the other method of cultivation, but where sites are of equal value the merits of the two systems may be judged on convenience and cost.

Shore-based systems can be reached only at low water whilst rafts are usually accessible at all times. Rafts are costly structures to build and maintain and this must be considered in relation to the value of their crop, which decreases relative to the carrying capacity of the raft as the spat grows. The selling price of oyster seed only approximately doubles for every tenfold increase in weight up to about 10 g live weight. Thus the value of a crop of 10 g oysters per unit area of raft is only one fifth of the value of 1 g oysters. It is unlikely, therefore, that raft culture could economically produce market-sized oysters, but it is satisfactory as a nursery for small oysters, which are eventually moved to shore-based sites for further on-growing.

Floating upwelling systems

These are specialised structures for growing nursery-sized oysters. Conventional trays are not ideal structures for holding oysters since they lie in the same plane as most tidal currents and, therefore, do not encourage an efficient exchange of water. One method of improving water flow through trays uses a deflector plate situated at the bottom of a floating, moored box (Fig. 4.8). Water is driven through the throat of the box at its base, passing upwards through the stack of trays within the box before escaping through the top rear sections at the water surface. This floating upwelling system (flupsy) showed good results in the Menai Strait, where it was moored in a channel where current speed reached 1–2 knots. The growth of Pacific oysters in a flupsy was greater than that in intertidal trays (Fig. 4.9).

Prolonged retention of stock in the same position within the stack leads to uneven growth (Table 4.1) because those nearer to the water intake receive more food at the expense of those higher up. Good management practice should ensure that the stock is evenly redistributed during regular servicing.

The commercial application of the method was demonstrated with a unit of eight flupsys (Fig. 4.10) built onto a 5 m × 5 m raft moored in an Essex

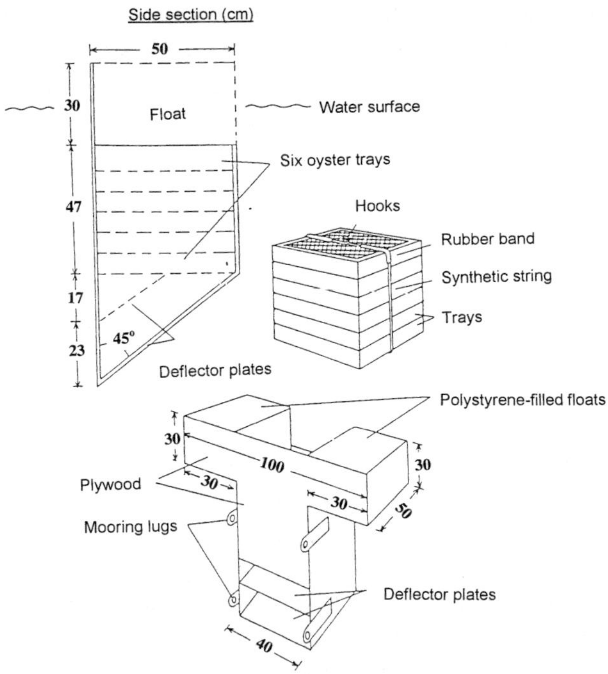

Fig. 4.8 Design of an experimental plywood floating upwelling system (Spencer & Hepper, 1981).

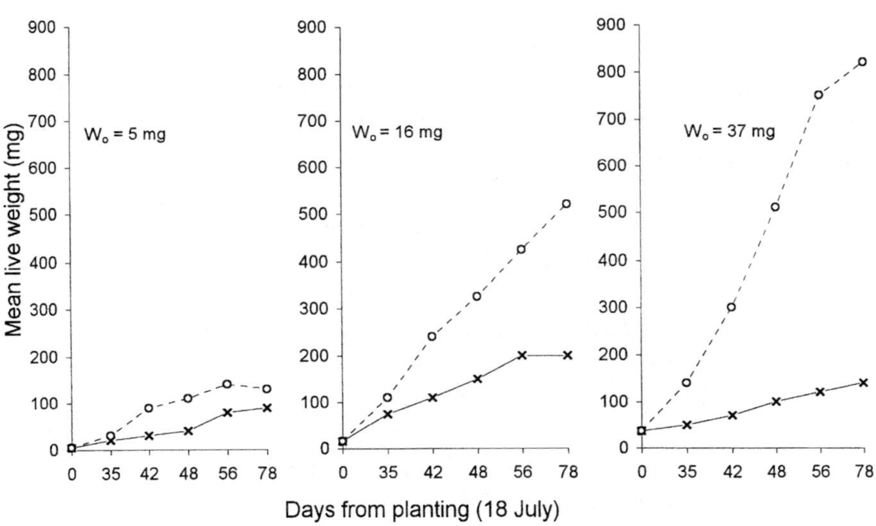

Fig. 4.9 Growth of three batches of Pacific oysters in a floating upwelling system (O) and in intertidal trays (X) in the Menai Strait in 1978.

Table 4.1 Menai Strait, Wales: Growth of 20 mg Pacific oysters after 3 months in six trays in tide-powered upwelling systems stocked (a, b) at two densities and (c) in three shore trays attached to a trestle (Spencer & Hepper, 1981)

	Mean live weight (mg)		
	Upweller (a) 1440 per tray	Upweller (b) 8660 per tray	Trestle (c) 1440 per tray
Tray 1 (top)	535	120	732, 613, 579
Tray 2	632	87	
Tray 3	763	84	
Tray 4	942	135	
Tray 5	948	288	
Tray 6 (bottom)	776	531	
Mean weight (% survival)	777 (78)	206 (84)	638 (62)

Water entered bottom of upwellers below tray 6 and left at the top, above tray 1

Fig. 4.10 Commercial flupsy in an Essex creek.

creek (Peter French Oysters Farms). This was used to grow 200 000 flat oyster spat, 10 mm in size, to a mean weight of 5 g (35 mm), a convenient size for on-growing in trays on trestles. Their initial bulk weight of 32 kg increased to 715 kg from June to October 1984, with 72% survival. At the end of the trial, the stock was housed in 160 trays (1 m × 0.5 m × 5 cm deep) at a density of 0.9 g/cm^2 of tray area. The weight of oysters that can be kept in these systems is dependent on the volume of water, and therefore of food, reaching the stock. Stocking density on the commercial raft reached 90 kg per flupsy but it is not known whether this was near to its maximum capacity. In contrast, the maximum capacity of the small experimental flupsy (Fig. 4.8) was 10 kg, a biomass at which no further growth was possible at its locality in the Menai Strait. With experience, the operator should be able to determine maximum carrying capacity of his flupsy at its site. Thereafter, he should avoid densities exceeding about 50% of the maximum capacity to ensure good growth and survival of the stock.

Careful site selection and management of these systems is required. They need to be sited in a tidal stream with as little slack water as possible at high and low tides to ensure that the stock are not unduly stressed.

Despite the faster growing conditions that these systems offered, they were not widely used commercially, possibly because of a lack of suitable sites and the intensive management needed.

Land-based and floating pumped upwelling systems

These are now used in hatcheries and nurseries for growing small bivalves to 5–10 mm, a size at which they can readily withstand the rigours of planting in the sea. It is an intensive method of cultivation whereby the stock is kept in relatively small containers (Fig. 4.11). Investment in trays and associated equipment is reduced, but to a certain extent this is offset by the high cost of pumping. Impounded water in disused quarries, gravel pits, or purpose-built ponds may be used to supply a unit ashore or on a raft. The raft may also be moored in a sheltered estuary at a site with an accessible electricity supply. One company dispensed with pumped water in favour of an air-lift, powered by compressed air supplied at low pressure.

Upwelling containers are usually PVC tubes sealed at the bottom with plastic mesh to retain the oysters. The container's diameter and height depends on the capacity required. A container measuring 30 cm diameter by 60 cm high holds a maximum of 5 kg of spat. When fitted with a 32 mm overflow and valve and with a 20 cm head of water, a water flow of 125 l/minute is possible at a specific flow rate of 25 ml/minute/gram live weight for 5 kg of spat. However, with narrower cylinders at high flows, water velocity may exceed 2–3 cm/second causing spat to lift from the bottom into the

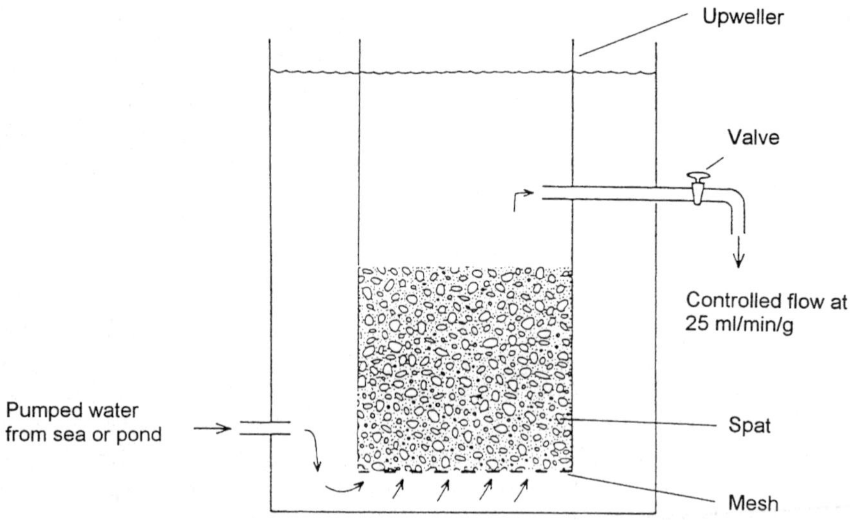

Fig. 4.11 Diagrammatic view of a pumped upwelling system.

overflow pipe. It is then necessary to reduce the biomass of spat while retaining the specific flow rate.

Although a specific flow rate of 25 ml/minute/gram provides good growing conditions in the summer, the flow rate for best growth is dependent on the food content of the water. This interaction between flow rate and food content is an important aspect of the management of the upwelling system. Observations on experimental upwelling nurseries indicate that optimum uptake of food by juvenile oysters and clams occurs at flow rates adjusted to give 20% filtration of the particles flowing through an upweller (Spencer, 1988). When food is abundant, low flow rates (≅20 ml/minute/g) are required, but when it is scarce high flow rates (≅50 ml/minute/g) are required to achieve 20% filtration and, therefore, the best growth. Automated control of flow rate to optimise growth of spat may be a possibility. This could be achieved using electronic sensors to measure continually the quantity of particles in the incoming and outgoing water. A central processor that controls the adjustment of water flow rate via an automatic valve to achieve 20% filtration could assess the information.

A unit producing a million spat at 0.2 g mean live weight (10 mm) would comprise one hundred upwellers (60 cm × 30 cm diameter) with a flow requirement of 12.5 m^3 of water per minute, at a maximum stocking density to ensure good growing conditions in the summer. The time required to reach this size depends on initial spat size, and temperature and productivity of the water, and is likely to range from 4 to 8 weeks for small spat (0.01 g, 3–4 mm) in the warmest months.

Fig. 4.12 Stanway cylinders attached to oyster racks (Port Stephens, New South Wales).

Other systems

Bivalves may be grown in containers of infinite variety in design so long as the basic requirements of good water exchange, exclusion of predators and low cost are met. Two systems gaining popularity in Australia are worthy of mention. These are the Stanway cylinder and the BST oyster longline, both of which encourage good water flow through the containers and, therefore, good growth and survival of the Sydney rock oyster (*Saccostrea commercialis*) and Pacific oyster for which they are used.

The Stanway cylinder (Fig. 4.12) is a cylinder of plastic mesh (1 m long × 0.25 m diameter; 5 mm mesh) with each of its two end walls moulded into a buoyant, flotation compartment. The cylinder is suspended horizontally on an external, rigid PVC axle-rod, which passes through a lug on each end wall and is fixed to a wooden frame so that it sits about 1 m above the intertidal beach. At low tide the cylinder hangs from the axle, but as the tide rises its buoyancy lifts and rotates the cylinder about its axis. Wave action and current movement ensures and that the oyster spat are gently tumbled about within the cylinder, regularly dispersing them and preventing them from smothering each other. The almost continuous movement of the spat against the mesh also minimises siltation and fouling. The cylinders are set up about 1 m apart, in rows, in the sheltered oyster fisheries of New South Wales, Australia, where tidal rise and fall is only about 1 m. As with all systems, good management of the cylinders is of paramount importance. The manufacturers recommend that optimum stocking density is 0.5–1 l of spat per cylinder. With volumes less than this, the cylinder has too much buoyancy for its load and the extra wave-induced movement damages the oyster spat, while with greater volumes water penetration is insufficient to support good growth and survival. Although the cylinders are primarily used for spat on-growing, the manufacturers recommend that market oysters are tumbled for the last 3–6 months before harvest to harden the shell and improve the meat and also to prepare the oyster for the shock of transport.

The BST longline system

This system, developed for Pacific oyster culture in the rough waters of South Australia is now being applied to the relatively calm waters of NSW oyster fisheries. The longline comprising 4 or 5 mm plastic wire is suspended from wooden posts, 3 m apart, sunk into the intertidal beach to leave 1–2 m above the ground. The oysters are held in plastic mesh cylinders shaped like an elongated pyramid, which are clipped to, and suspended from, the wire at 3 units per pair of posts and *c.* 0.5 m above the ground. The system is designed for growing fairly large oysters to market size over a 2½ year period. The oysters, stocked initially in 8 mm mesh cylinders, are graded every 6 months and are moved progressively onto larger meshes of 12 mm, 17 mm and 19 mm mesh size.

Husbandry

The options available for growing seed oysters to market size (Fig. 4.3) range from permanent confinement of the stock in trays throughout cultivation to temporary confinement until the stock reaches an intermediate size at which it can be transferred safely to ground layings. The cultivator's preferred method should reflect the suitability of the area for ground cultivation and the extra cost of growing larger oysters in trays.

This section emphasises the main requirements for growing Pacific oyster seed in trays to 10 g live weight and thereafter to market size on intertidal ground lays. The appendix at the end of this chapter describes a procedure for handling spat on receipt from the nursery supplier.

Tray culture

Stocking density trials with Pacific oysters in the Menai Strait and elsewhere in England and Wales, using trays of various types and mesh sizes, have shown that high stocking levels cause poor growth, clumping (i.e. the fusion of two or more oysters) and high mortalities. The general conclusion is that during the first two growing seasons a density of about 1 g live weight of oysters

Fig. 4.13 Four sizes of Pacific oysters in NWP trays at 0.5 g/cm^2 (top left clockwise: mean live weights 0.75 g, 2.0 g, 4.0 g, 6.7 g).

per square centimetre of tray bottom (1 g/cm^2) is acceptable for short periods (1 month). Densities of 2 g/cm^2 and above, however, are acutely stressful to the oysters and should be avoided. A good practice is to stock oysters at levels which do not exceed 0.5–1 g/cm^2 in trays with meshes of 5 mm or larger. Four sizes of Pacific oysters in North West Plastics (NWP) trays (50 cm × 50 cm × 5 cm deep; 2500 cm^2 in area) at 0.5 g/cm^2, are shown in Fig. 4.13. Oysters too small for 5 mm mesh are stocked at a lower level (2.5 oysters per cm^2); 0.02–0.2 g/cm^2) to ensure a good start to their new life in the sea.

In winter, when growth is very slight, oysters can be stocked at densities of up to 2 g/cm^2 because their food and water requirements are very much reduced. However, silt deposition is usually greater in winter and regular servicing may be required to prevent smothering.

With tray culture, oyster growth in relation to size and temperature requires careful consideration. Oyster size and sea water temperature have a large and fairly predictable effect on growth. Knowledge of the relationship between these two factors presents the cultivator with the opportunity to manage tray stocking levels to achieve good growth and survival. Growth data shown in Fig. 4.14(a), derived from observations at various sites in England and Wales, have been used to calculate the tray stocking density

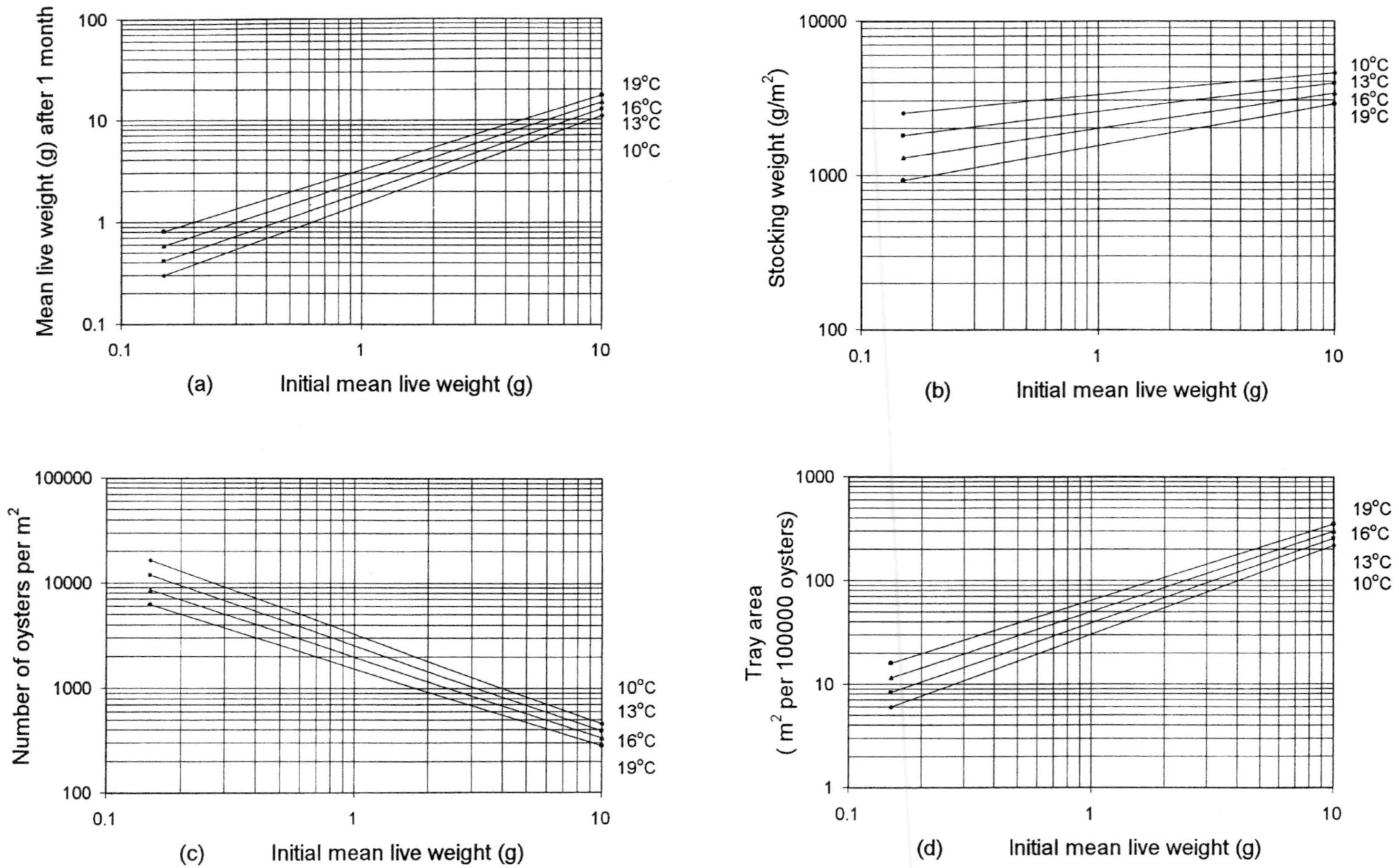

Fig. 4.14 Pacific oysters: calculated growth and tray stocking density required to achieve a target density of 0.5 g live weight of oysters per cm^2 of tray area after 1 month in the sea. From Spencer & Gough 1978, Spencer *et al.*, 1985.

requirements for Pacific oysters to achieve a target density of 0.5 g/cm^2 after 1 month in the sea. These data are expressed in various ways as an aid to tray management and are shown in Figs 4.14(b–d). Options such as raising the target density to 1 g/cm^2, allow stocking biomass to be increased approximately by the same proportion. Similarly, increasing the interval between servicing requires the lowering of the stock weight to a level commensurate with the length of interval and the sea temperature during that interval.

To use information in Fig. 4.13 for the management of tray density, knowledge of sea water temperature at the site is required. The mean monthly values for various coastal sites in the UK (see Chapter 11, Table 11.2) provide some idea of seasonal and geographical variation around the country and may be used in the absence of specific information for the site.

Examples of the initial stocking densities required for 0.3 g (15 mm) Pacific oysters held in a NWP tray to yield 0.5 g/cm^2 after one month at three temperatures are shown in Table 4.2.

Frequency of servicing and adjustment of stocking density are important factors that contribute to the welfare of oysters in trays. In the first year, with small mesh trays, frequent servicing and adjustment of stocking density are required. Although re-grading monthly is cost efficient in the use of trays, it is labour intensive. Longer intervals between re-grading may be possible depending on the rate of silting and fouling within the tray. From the second year onwards, frequent and excessively rough handling retards growth. It is then beneficial to limit re-grading to intervals of 3–6 months but with regular checks to ensure control of silting and fouling. No matter how uniform the initial size of the oysters is, individuals grow at different rates to produce a wide range of sizes after a few months. A typical distribution of sizes within a population, by the percentage and mean weight of oysters retained by sieves of different mesh sizes in successive months is given in Table 4.3.

Several hundred thousand spat of less than 1 g can be graded quickly by one person using a hand sieve (0.25–0.5 m^2 surface area) shaken gently in a

Table 4.2 Examples of initial stocking densities of Pacific oysters in trays

Expected mean monthly sea temperature (°C)	Initial live weight (g)	Predicted live weight (g)	Initial density per NWP tray	
			(a) weight (g)	(b) number
10	0.3	0.55	680	2220
16	0.3	1.0	375	1250
19	0.3	1.3	290	960

Table 4.3 Pacific oysters in trays in the Menai Strait, showing growth as percentages of the population retained by sieves of different mesh sizes. The mean live weights (g) are shown in parenthesis

Month	Mean sea temperature (°C)	Sieve mesh size (mm) 2	7	11	18	25
April	7.5	100 (0.01)				
May	11.4	100 (0.02)				
June	13.4	83 (0.06)	17 (0.12)			
July	15.3	16 (0.10)	65 (0.21)	19 (0.40)		
August	15.8	1 (0.19)	78 (1.5)	21 (3.0)		
September	15.5		7 (0.6)	48 (1.8)	25 (3.7)	20 (6.6)

Table 4.4 Optimum stocking densities of oyster seed in different systems

System		Stocking density (g/m^2)	Author
Trays on raft or shore	Current speeds of 50–100 cm/sec	0.5–1.0	Spencer *et al.*, 1978, 1985
Trays on a tide-powered Hepper, upweller	"	6	Spencer & 1981
Pumped upweller container	Outdoor nursery using fertilised water	10	Spencer, 1987

trough of water, but larger numbers and larger oysters require mechanical grading. A range of hand sieves of 2–25 mm mesh is required for dealing with spat up to 5 g.

Stocking density of systems

The number or weight of oysters that may be stocked per unit area of tray is very much influenced by the amount of food that reaches the oysters. Trays suspended from a raft or on an intertidal trestle are not best placed for an efficient exchange of water during normal tidal flow. Extra water flow induced by deflector plates on tide-powered floating upwelling systems or on land-based pumped upwelling systems enable higher stocking densities to be employed (Table 4.4).

Intertidal trays, however, are the cheapest option and can be used for growing oysters to market size. Tide-powered and pumped upwelling systems are relatively expensive systems to employ and become uneconomical once spat reach *c.* 1 g in size.

Intertidal ground layings

Experiments to determine the most suitable stocking density for Pacific oysters re-laid intertidally on the ground have been made for 1 g oysters within a protective fence and for around 10 g oysters in unprotected pens. One-metre square plots were established on soft, muddy ground at 10% tidal exposure in the Menai Strait and growth and survival were recorded over a period of time in the following way:

(1) The growth of small oysters within a protective fence was tested by re-laying 1.3 g oysters at densities ranging from 160 to 2000 per m^2 (Table 4.5). After 3 months no differences were observed in the mean weights of the oysters at densities up to 640 per m^2; 50% or more of the individuals were larger than 10 g and, therefore, of crab-proof size. Oysters re-laid at 1040 and 2000 per m^2 were smaller than those re-laid at lower densities. After a further season's growth, the relationship was even more distinct. Those initially at 160 and 240 per m^2 were of similar weight, but at higher densities the mean weight and proportion of marketable oysters decreased with increasing density. The practical significance of this is that stocking densities of 500–600 per m^2 can be employed for growing stock to about 10 g but a lower level, of about 200 per m^2, is required in growing them to market size. Thinning, however, requires considerable effort and it may be more cost effective to stock

Table 4.5 The effect of stocking density on the growth and survival of 1.3 g Pacific oysters, re-laid intertidally on the ground inside a protective fence in August 1977

	November 1977				October 1978			
Initial number per m^2	Mean weight (g)	Total weight (kg/m^2)	Percentage >10 g	Survival (%)	Mean weight (g)	Total weight (kg/m^2)	Percentage >50 g	Survival (%)
160	11.8	1.6	54	86	57.6	6.7	72	72
240	11.9	2.5	58	87	58.9	8.6	66	61
400	11.4	3.8	68	84	53.1	13.5	48	64
640	10.9	6.0	54	86	47.6	17.3	41	57
1040	7.0	5.4	21	74	40.6	15.7	10	37
2000	6.5	9.6	10	74	34.4	19.8	17	29

Table 4.6 The effect of stocking density on the growth and survival of 8.6 g Pacific oysters re-laid intertidally on unprotected ground layings in April 1977

	December 1977				October 1978			
Initial number per m^2	Mean weight (g)	Total weight (kg/m^2)	Percentage >50 g	Survival (%)	Mean weight (g)	Total weight (kg/m^2)	Percentage >100 g	Survival (%)
50	58.6	2.5	66	80	129.2	5.0	66	78
100	61.4	4.4	NR	72	128.4	8.8	83	68
150	61.2	6.9	NR	71	128.0	13.3	78	69
200	61.6	8.2	78	67	127.8	16.2	79	64

initially at 200 per m^2 and leave them undisturbed until they reach market size.

(2) The growth of large oysters was tested by re-laying 8.6 g (44 mm) oysters in unprotected pens at densities ranging from 50 to 200 per m^2. The trial was started in April and growth and survival were recorded at the end of two growing seasons (Table 4.6). More than 60% of the population at each density reached minimum market size (50 g) in 8 months and maximum market size (100 g) in 16 months. Stocking density within the range of 50–200 per m^2 did not seriously affect growth and survival, which confirmed the conclusion drawn from Table 4.5 that oysters can be grown to market size at about 200 per m^2.

(3) Annual growth of Pacific oysters of various sizes, re-laid at the recommended density of 200 per m^2 on ground layings in the Menai Strait is shown in Fig. 4.15. The survival of these oysters after 2 years (Fig. 4.16) is very sensitive to their size at re-laying. These data show that 10 g oysters would be expected to grow to about 70 g in 2 years, with an average survival of 50%.

Costs of cultivation

The hatchery prices of Pacific oysters of various sizes range from £7 to £30 per 1000 for stock 3–30 mm in size (1999 prices). The value of market-sized oysters is variable depending on the outlet but may reach £200 or more per 1000 oysters at first sale.

The cost of growing 3–4 mm Pacific oysters to market size was assessed in costed trials in the Menai Strait. Two strategies were adopted in two 4-year trials. In the first trial, 166 000 seed were grown in trays for 1–1½ years and then on intertidal ground lays to market size. In the second trial, 62 000 seed were grown to market size solely in trays. Little difference in the final mean sizes of the oysters grown by the different strategies (85 g and 80 g)

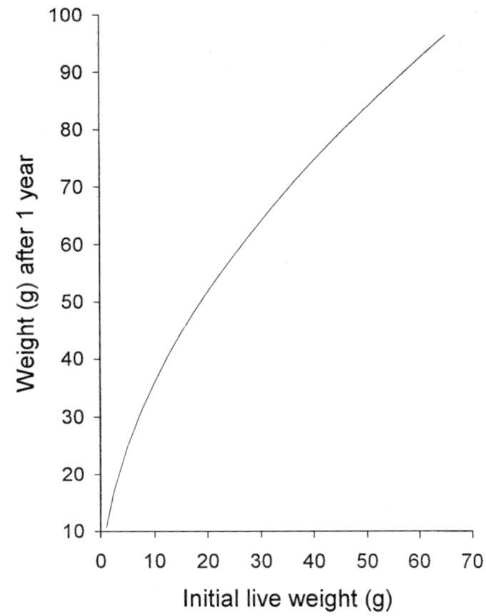

Fig. 4.15 Annual growth of Pacific oysters, stocked initially at 200 per cm^2, on intertidal ground layings in the Menai Strait.

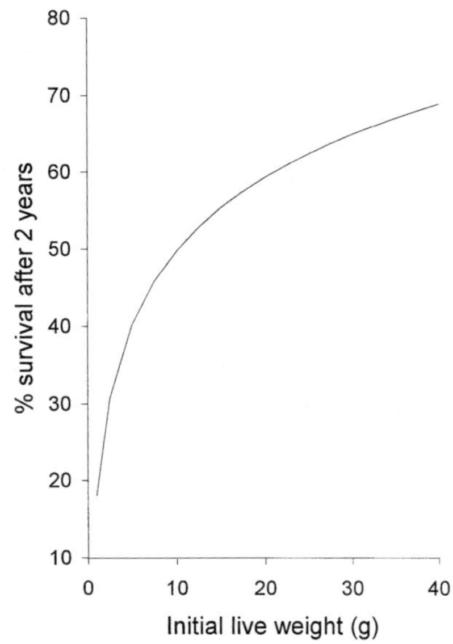

Fig. 4.16 Survival of Pacific oysters after 2 years on intertidal ground layings in the Menai Strait.

was found but survival was twice as high for oysters grown solely in trays. Since the costs of rearing the oysters as a proportion of their final values were similar between strategies, profitability was related to survival. Thus, even though tray culture is relatively expensive, improved survival should provide a better yield in net income, in this instance of about twice that of ground culture (Table 4.7).

Table 4.7 Some statistics of growing Pacific oysters to market size in costed trials by (a) tray and ground culture and (b) tray culture alone

	(a)	(b)
% survival	22	48
Final mean live weight (g)	85	80
Costs as % of final value	56	52
Number of spat needed per 1000 marketed oysters	4700	2100
Net income (£ per 100 000 seed; 1999 prices)	3500★	7500★

★ calculated from original 1982 data (×2.17) to allow for change in purchasing power of £

Appendix: Procedure for handling and transporting nursery spat into trays

The following example shows the necessary steps for accommodating 50 000 Pacific oyster spat (mean shell length 9 mm, and mean live weight 0.15 g) in NWP trays in the Menai Strait in May.

(1) Transportation: the commercial nursery will transport the seed quickly to its destination (usually within 24 hours by post). The seed are normally packed (damp) in a sealed plastic bag, within an expanded polystyrene box to protect them against physical damage, drying-out and temperature extremes. The spat must be kept cool and shaded since exposure to sunlight causes overheating and loss of water from the shell cavity.

(2) Estimation of numbers, mean weight and stocking density: on receipt, the spat should be placed in the sea in trays at the appropriate density.
   (a) The number and mean live weight of oysters is usually provided by the nursery, or it may be obtained as follows:
      (i) weigh spat and container.
      (ii) weigh container.
      (iii) weigh and count several samples of about 500 oysters. Suitable and cheap spring balances are available. Two with capacities of 25 kg (×100 g) and 1 kg (×10 g) are required.
      (iv) calculate the number of oysters by:

$$\frac{\text{Total weight} - \text{Container weight}}{\text{Weights of samples} - \text{Weights of containers}} \times \text{Number in samples}$$

      (v) calculate the mean weight of oysters by:

$$\frac{\text{Weight of samples} - \text{Weights of containers}}{\text{Number in samples}}$$

(b) Tray stocking density can be determined as follows:
  (i) From Table 5.2 (Chapter 5), it can be seen that the mean sea water temperature for the Menai Strait in May is about 11°C.
  (ii) From Fig. 4.14, it can be seen that the stocking weight per NWP tray (i.e. 0.25 m^2 in area) for 0.15 g oysters at 11°C is *c.* 550–600 g and that the area of trays required for 50 000 oysters is *c.* 3.5 m^2, which is equivalent to 14 NWP trays.

(c) The trays are now stocked at the required density and should be placed in the sea as soon as possible.

(3) Oyster growth and further holding requirements: The on-grower may plan his further tray requirements 1 month ahead, by estimating the expected size which his stock will reach at the end of May (i.e. 0.32 g from Fig. 4.14) and repeating the procedure in (b) above for a temperature of 14°C (i.e. expected temperature for June in the Menai Strait from Table 5.2, Chapter 5).

References and bibliography

Dare, P. J., Davies, G. & Edwards, D. B. (1983) *Predation on Juvenile Pacific Oysters* (Crassostrea gigas *Thunberg) and Mussels* (Mytilus edulis *L) by Shore Crabs* (Carcinus maenas (*L.*)). Fisheries Research Technical Report, MAFF Directorate of Fisheries Research, Lowestoft, **73**.

Davies, G., Dare, P. J. & Edwards, D. B. (1980) *Fenced Enclosures for the Protection of Seed Mussels* (Mytilus edulis *L) from Predation by Shore Crabs* (Carcinus maenas (*L.*)). MAFF, Fisheries Research Technical Report, MAFF Directorate of Fisheries Research, Lowestoft, **56**.

Spencer, B. E. (1987) *Bivalve Upwelling Culture: a Review of Progress*. MAFF, Fisheries Laboratory Internal Report.

Spencer, B. E. (1988) Growth and filtration of juvenile oysters in experimental outdoor pumped upwelling systems. *Aquaculture*, **75**, 139–58.

Spencer, B. E. (1990) *Cultivation of Pacific oysters*. Laboratory Leaflet, MAFF Directorate of Fisheries Research, Lowestoft, 63.

Spencer, B. E. & Gough, C. J. (1978) The growth and survival of experimental batches of hatchery-reared spat of *Ostrea edulis* L. and *Crassostrea gigas* Thunberg, using different methods of tray cultivation. *Aquaculture*, **13**, 293–312.

Spencer, B. E. & Hepper, B. T. (1981) Tide-powered upwelling systems for growing nursery-size bivalves in the sea. In: *Nursery Culturing of Bivalve Molluscs* (eds C. Claus, N. de Pauw & E. Jaspers), pp. 283–309. Special Publication of the European Mariculture Society, Bredene, Belgium, **7**.

Spencer, B. E., Gough, C. J. & Thomas, M. J. (1985) A strategy for growing hatchery-reared Pacific oysters (*Crassostrea gigas* Thunberg) to market size – experiments and observations on costed small-scale trials. *Aquaculture*, **50**, 175–92.

Williams, P. (1981) Offshore nursery culture using the upwelling principle. In: *Nursery Culturing of Bivalve Molluscs* (eds C. Claus, N. de Pauw & E. Jaspers), pp. 311–15. Special Publication of the European Mariculture Society, Bredene, Belgium, **7**.

5 Clam cultivation

Introduction

Clams, including cockles and arkshells, are the second major contributor to the world catch of molluscs with about 17% of the recorded landings. FAO statistics record landings of 2.8 million tonnes in 1997 (Table 5.1; FAO, 1999).

Most of the catch is derived from harvesting natural populations. Many stocks are managed to ensure that they are not over-exploited. Regulatory devices usually restrict fishing effort by the issue of licences and by controlling gear, season of fishing, size of catch, minimum landing size, etc. The cockle fisheries of the Netherlands and UK are good examples of regulated fisheries. In some parts of the world, e.g. China (razor and Manila clams), USA (soft and hard shell clams), Spain and Portugal (grooved carpet shells), wild seed may be collected and transferred to parks for cultivation. The proportion of world production attributable to cultivation, however, is not precisely known or easily derived. Many countries do not report whether landings are from cultivated or unmanaged wild stocks, a situation compounded by the grey area of definition between the two categories.

On a world scale, the Manila clam is the most important species with 48% of the landings, mainly from Japan, China and the Republic of Korea (Table 5.1). Next in importance are the blood cockles and razor clams from the Far East, which together make up 22% of the landings.

Cultivated wild seed

In some countries abundant wild seed are often harvested for further cultivation. Usually the technology for seed collection is fairly simple with hand tools and sieves frequently used. In Maine, USA, soft shell clam seed are sometimes gathered by digging, but hydraulic dredges have been developed which permit collection from areas with dense settlements. The seed are then re-laid on culture plots at more productive densities (Hidu & Newell, 1989;

Table 5.1 Main areas of clam production (FAO, 1999)

Species	Common name	Country	1987	1997	10 year average*
			Thousands of tonnes		
Anadara spp.	Blood cockle (ark shells)	Far East (mainly China, Korea, Malaysia, Indonesia and Thailand)	146.0	251.6	200.7 (125%)
Arctica islandica	Ocean quahog	Mainly USA	187.6	163.8	184.1 (89%)
Cerastoderma edule	Common cockle	Europe (mainly The Netherlands and UK)	107.8	40.9	82.4 (50%)
Spisula solidissima	Surf clam	Mainly USA	149.2	140.6	164.1 (86%)
Chamelea (Venus) gallina	Striped Venus	Europe (mainly Italy and Turkey)	43.6	37.7	46.8 (81%)
Tapes (Ruditapes) decussatus	Grooved carpet shell	Europe (mainly Spain, Portugal, Tunisia and France)	11.9	5.6	12.8 (44%)
Tapes (Ruditapes) philippinarum	Manila clam/Japanese littleneck clam	China, Japan and Korea	297.6	1331.6	703.2 (190%)
Protothaca thaca	Taca clam	Chile	35.0	12.5	28.4 (44%)
Mercenaria mercenaria	Hard shell clam/ northern quahog	Mainly USA and Taiwan	37.2	49.9	46.2 (108%)
Solen/ Sinonovacula spp.	Razor clam	Mainly China	134.3	354.6	205.6 (172%)
Mya arenaria	Soft shell clam	Canada and USA	21.9	6.8	13.1 (52%)
	World total all species		1554	2758	2046 (135%)

(*) 1997 values as % of 10 year average (1987–1996) are shown in parenthesis

Newell, 1991). Increased production, however, requires the freshly re-laid seed to be protected from predation.

China, perhaps more than any other country, has exploited the collection and cultivation of wild clam seed. Three species, razor clams, blood cockles and Manila clams are especially targeted.

Razor clams

Intertidal culture plots are prepared before the expected spatfall by loosening and smoothing the substrate. Careful monitoring of the abundance and development of pelagic larvae in the water is undertaken to forewarn the fishermen of impending spatfalls. Within 3–5 months after spatfall, the seed, *c.* 10 mm in size, are transplanted to other prepared beds at densities of 900–1800 per m^2. Market size of *c.* 50 mm is reached about 6 months later with an average yield of 3 kg/m^2 (Nie, 1991).

Blood cockles (ark shells)

Shallow ponds are used for the settlement and initial growing of seed. Dense spatfalls require thinning several times and surplus seed are placed on intertidal beds. The cockles are harvested after 2 years when they are *c.* 20 mm in size with yields of 2–6 kg/m^2 (Nie, 1991).

Manila clams

Seed are also collected in shallow ponds, several hectares in area located low on the intertidal zone. The ponds may be inoculated with *Chaetoceros* as food for the larvae and spat. Tidal exchange is managed to prevent the food resource from being substantially flushed out. The ponds can produce several batches of spat per year with yields of 750–1500 per m^2 of 5 mm clams. The seed are re-laid at densities of *c.* 180 per m^2 on prepared intertidal beds, reaching market size of 35 mm after a further year. Yields of 2–4 kg/m^2 are obtained (Nie, 1991).

Like the Pacific oyster, and within the same time period of the twentieth century, the range of the Manila clam has been widened beyond its natural Far East habitat by accidental and intentional introductions. It reached the western seaboard of North America accidentally from Japan with importations of Pacific oysters into British Columbia, Canada, and Washington State and California, USA, in the 1930s (Bourne, 1989; Chew, 1989). It has become naturalised in these areas, occupying the intertidal zone and forming economically important wild and recreational fisheries in British Columbia and Washington State. Its introduction into Europe was intentional, with broodstock imported from Washington State into France in 1972 (Goulletquer & Héral, 1997) and Wales in 1980 (Utting & Spencer, 1992) to furnish these countries' home and overseas hatchery trades. Re-laid hatchery-reared seed have led to natural spatfalls in some areas of Europe. These have led to a small fishery in south Brittany, France (1000–1500 t/year in 1989 and 1990) and Poole Harbour, England (*c.* 50 t/year in 1997), and a substantial fishery in the Venice lagoons and Po delta, Italy, reaching 27 000 t in 1991 (Mattei & Pellizzato, 1997).

Growing hatchery-reared seed

Clams generally, are easy to rear in the hatchery (see Loosanoff & Davis, 1964) but only a few species are produced commercially in any significant numbers by this method. Hatchery methods are well known, reliable and widely available, especially in countries where natural clam seed production is unreliable or insufficient to satisfy demand. In North America, *Mercenaria mercenaria*, *Tapes philippinarum* and *Mya arenaria* are probably the commonest species produced in commercial hatcheries. In France, Spain and England, *T. philippinarum* and *T. decussatus* are commonly produced and *M. mercenaria* occasionally. Husbandry requirements for growing hatchery-reared clams in the sea are generally similar between species, and internationally follow the same general pattern (see Anderson *et al.*, 1982; Anon., 1988).

The following account describes experimental work aimed at determining the requirements for growing hatchery-reared Manila clams and palourdes in the sea in England and Wales (Spencer *et al.*, 1991).

The Manila clam was introduced into the UK (via the hatchery quarantine facilities of the Fisheries Laboratory, Conwy) because of its reported fast growth rate and lucrative markets on the continent of Europe with high market values in France, Spain and Italy. Trials confirmed that the species does grow quickly, reaching a minimum market size of 20 g (*c.* 45 mm) in 2–3 years. This is about 1 year less than for the palourde.

The Manila clam is superficially similar in shape and size to the palourde, but has distinctive black and white shell markings. There are other noticeable differences between the two species. The orange colour of the flesh, especially the foot, of the Manila clam contrasts with the off-white colour of the flesh of the palourde. Also, the siphons, seen when the clams are immersed in the water and pumping actively, are joined in the Manila clam but are separate in the palourde (Fig. 5.1).

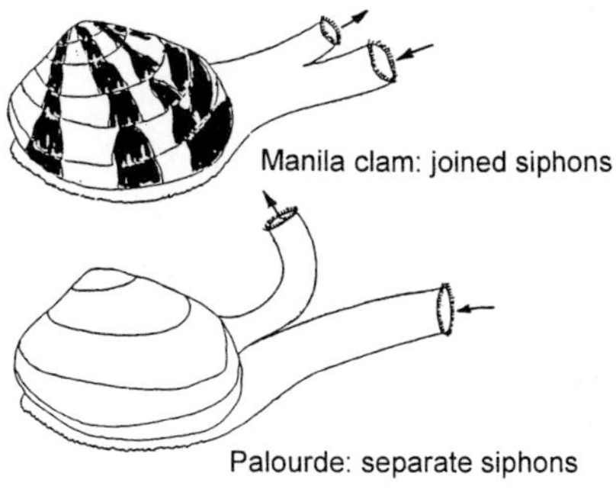

Fig. 5.1 Diagram showing the difference between the siphons of the Manila clam and the palourde.

Hatchery seed

Since Manila clams do not breed successfully in British waters, the on-grower is dependent on a commercial hatchery for his supply of seed. The seed is available at various sizes, ranging from *c.* 5 to 30 mm shell length (costing from £7 to £30 per thousand; 1999 UK commercial hatchery prices). The relationship between shell length and live weight is shown in Fig. 5.2.

Cultivation in the sea

Site

A site sheltered from extreme wind, wave and tidal action provides the best option for growing clam seed. More exposed locations may be suitable for cultivation, but greater effort and more robust equipment are usually required to manage the site effectively. The greatest problem with exposed sites is the likelihood of mobile sediment smothering net-covered clams before remedial action can be taken.

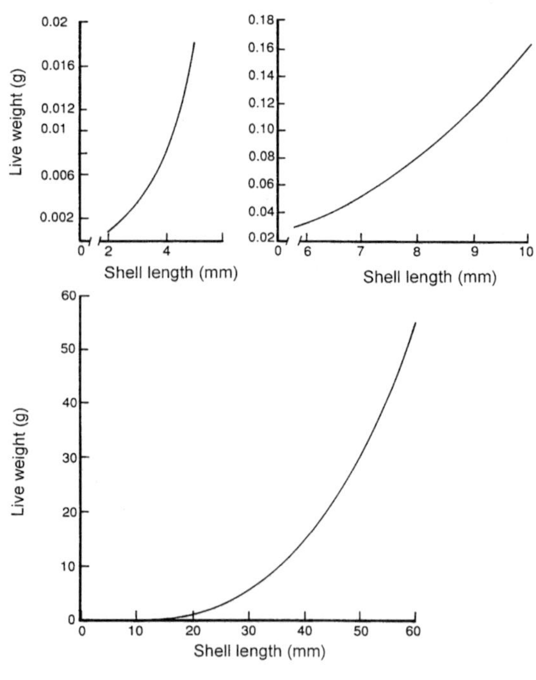

Fig. 5.2 Relationship between shell length and live weight of Manila clam.

Tray cultivation is likely to play a relatively small part in clam cultivation, since clams grow better in the soil. Selecting a site with a suitable substrate is very important. Clams grow well in a variety of sediments, from soft mud to firm sand and gravel. It is important, therefore, to consider access and the type of equipment likely to be used in planting, husbandry and harvesting. Some beaches may support wheeled or tracked vehicles, as used in France, which have access from the land. Some beaches are too soft or are inaccessible and require the use of a boat to transport equipment. Very soft mud and hard stony ground are generally impracticable, but their suitability may be improved by adding shell, sand or gravel. The soil should also be stable, since erosion or deposits of sediment on clam lays can cause loss of stock.

Manila clams and palourde prefer sea water with a salinity above 24 psu. Open coastal waters are usually between 30 and 35 psu, with little seasonal variation. Estuaries have a daily input of fresh water and, consequently, their salinity may vary between zero and 35 psu, depending on the proximity of the sea, the state of the tide and rainfall. Although laboratory trials show that Manila clams burrow actively at temperatures as low as 5°C, it is best to plant them during the growing season (April–October), when sea temperatures are usually above 10°C. Clams that have burrowed into the soil are quite tolerant to exposure to air at low tide. Although best growth is achieved in plots established near to the low mark of spring tides (LWST), good growth is still possible at higher levels of the beach.

Tray culture

Clams of *c.* 3–5 mm in size may be grown in trays with a mesh of appropriate size until they reach about 10 mm, which is an ideal size for planting in ground plots. Trays should be supported on trestles near to LWST. Commercial plastic trays measuring 50 × 50 × 5 cm with 5 mm mesh apertures (see Fig. 4.6, Chapter 4) are suitable for this first stage. They are rigid and present a flat base for even distribution of clams. Clams, when kept in bags or pôches, tend to crowd in segments or rucks caused by the sagging of the container. This problem may be partly alleviated by inserting a rigid PVC tube along each side of the bag or by using a bag with specially re-enforced sides.

Long-term culture of clams in trays encourages abnormal shell growth leading to a condition known as 'in-growing toe-nails'. This condition, also recorded in the common cockle, arises in the absence of support from the sediment, causing excessive gape of the valves, which leads to slow and abnormal growth at the shell margins. The incidence of this deformity may be sufficiently high to make the clams unmarketable.

Growth in trays

Large differences in growth may be expected between sites and between species, with the Manila clam growing up to twice the rate of palourde, depending on size, during one growing season (Fig. 5.3). Survival is also likely to be variable depending on husbandry and biological factors. The annual survival of the Manila clams shown in Fig. 5.3 for all trial sites averaged 65%.

Stocking density in trays

Since clams obtain their food by filtering microscopic plants and non-living organic particles from the sea water, it is important to avoid overstocking the trays. Competition for food causes reduced growth, and eventually high mortalities may occur in heavily overstocked trays. As yet, the optimum stocking density for tray cultivation has not been defined. In the absence of information specifically for clams, stocking rates for Pacific oysters may be used. Pacific oysters grow well at densities not exceeding 0.5–1.0 g of oysters per cm^2 of tray bottom. Thus a tray with an area of $0.25\,m^2$ has a maximum

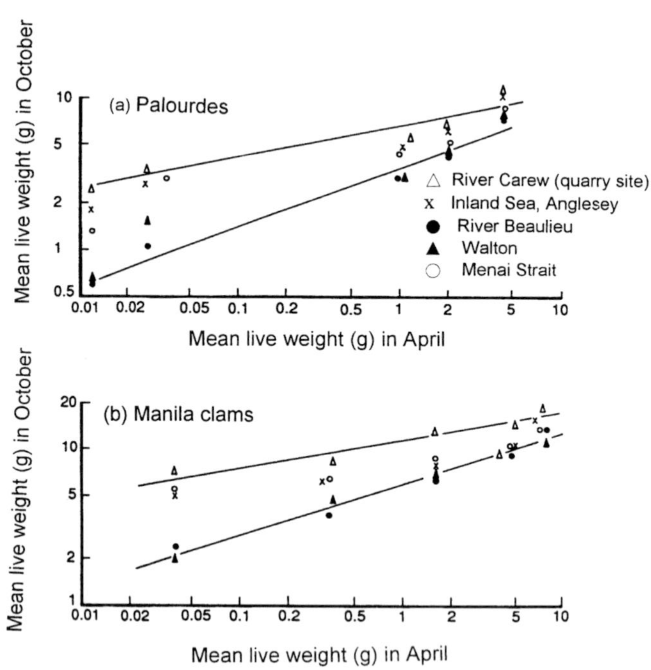

Fig. 5.3 One season's growth of palourdes and Manila clams in trays at five sites in England and Wales. Lines show range of growth between the best and poorest sites.

stocking capacity of 1250–2500 g. To ensure rapid growth during the summer, a tray should be stocked at about one quarter of this maximum to allow for growth 1 month later. A monthly adjustment of stocking density and rigid servicing schedule are required to maintain this stocking level and to ensure that clams are not overcrowded.

Ground culture

Since the sea bed is the natural habitat for clams, the trials described in this section have been carried out in ground plots. These were established intertidally, usually near to LWST, and consisted of areas of soil confined within a wooden frame (50 × 50 × 15 cm) pressed into, and with its surface flush with, the soil. The clams were protected from predation and outside interference by a covering of plastic netting with a mesh size appropriate to the size of the clams. Mesh sizes of 2.5 mm, 5 mm and 10 mm were used experimentally, of which the latter two are appropriate to commercial practice. The experimental plots were set up in six estuaries or sheltered waters (Menai Strait, Gwynedd; Walton-on-the-Naze and River Blackwater, Essex; River Beaulieu, Hampshire; River Exe, Devon; River Helford, Cornwall) in England and Wales. These trials examined growth and survival in relation to size at planting, stocking density, crab predation and exposure to air to determine the best cultivation practice.

Growth and survival in ground plots

Growth observations were taken over a number of years and for a range of sizes of Manila clams (Fig. 5.4). The observations were taken at the beginning and end of the growing season (March/April to October/November), thus effectively recording annual growth since little increase in size occurs outside of this period. These data have been used to predict growth at yearly intervals (Fig. 5.5). This is a useful guide in determining the length of time required for a Manila clam of a given size to reach market size. For example, a 10 mm (0.2 g) clam should grow to about 30 mm (6 g), 42 mm (18 g) and 51 mm (33 g) after 1, 2 and 3 years, respectively. A minimum market size of 20 g should be reached, on average, in the third growing season under these growing conditions.

Small clams survive less well than large clams in ground plots. The mean and range of survival of four size groups of Manila clams planted in April at four sites are shown in Table 5.2. Although on average the larger clams survived better than the smaller ones, it was apparent that the site could have a major effect on survival. In this trial, soil instability was a feature at two of the sites, and this factor, manifested by the deposition of thick layers of sediment on the clam plots, was responsible for high mortalities of clams in

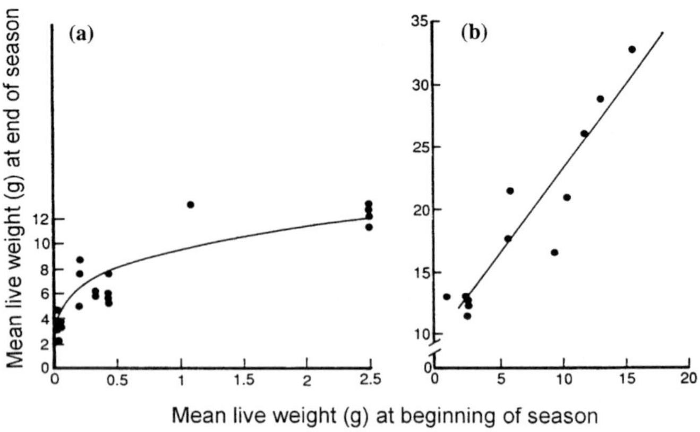

Fig. 5.4 One season's growth of (a) small and (b) large Manila clams in ground plots. Observations from several sites in England and Wales.

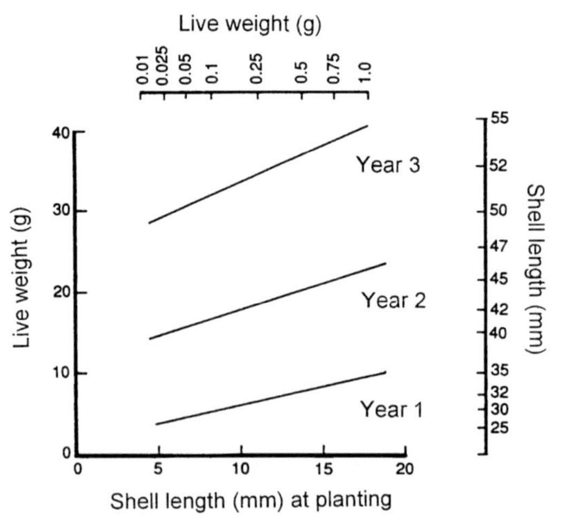

Fig. 5.5 Estimated growth of Manila clams after 1–3 years in the ground.

all size groups. Deposits of soil more than 1–2 cm deep can suffocate clams, since their siphons are unable to reach the water to obtain the oxygen necessary for survival.

The significance of differing size-related survival rates is of relevance to the economic benefit gained by purchasing a particular size of seed from the hatchery/nursery. The average or the best survivals shown in Table 5.2 suggest that there may be little benefit in buying the largest seed, since the gain in survival is outweighed by the high cost of the product. Other factors,

Table 5.2 Survival (%) of Manila clams after one year

Initial size		Survival (%)	
Shell length (mm)	Live weight (g)	Mean	Range between sites
3	0.01	34	0–58
4	0.02	47	2–78
10	0.2	60	0–88
24	2.5	77	37–93

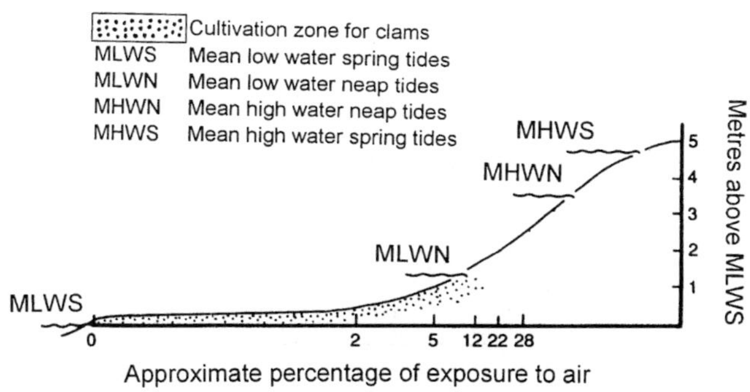

Fig. 5.6 Diagrammatic profile of the beach at the experimental clam site in the Menai Strait.

such as the extra labour and husbandry required to grow small clams to market size, need to be taken into consideration.

Growth and survival in relation to exposure to air

Gently sloping beaches, with an inclination less than 2%, usually provide wide areas suitable for cultivation of clams. Although Manila clams grow best near to LWST, reasonable growth may be obtained up to about low water of neap tides (LWNT), owing to their high tolerance of exposure to air. The relationship between growth and exposure to air (see Fig. 11.3, Chapter 11) shows that clams initially of 1 g live weight reached market size of 20 g in two years at levels of up to about 30% exposure to air. At the experimental site in the Menai Strait, this would be equivalent to a height of about 2 m above LWST (Fig. 5.6). Survival averaged 70% after 2 years and was not affected by exposure except at the highest station, at about 50% exposure, where total loss of clams, due to smothering by the mobile sediment of the upper beach, was recorded. The decline in live weight with increased exposure was matched by a decline in meat weight and shell weight. Since these decreased at about the same rate, there was no significant change in condi-

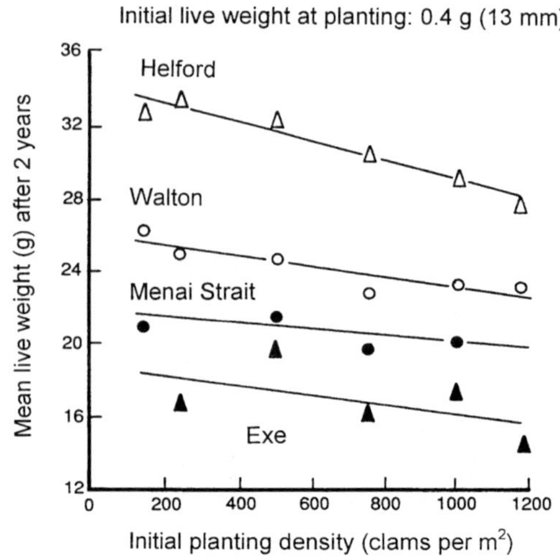

Fig. 5.7 Growth of Manila clams in ground plots in relation to stocking density.

tion index due to exposure to air. This situation is the same with Pacific oysters but, whereas oysters in trays stop growing at 30–35% exposure, clams at these levels in the ground still maintain some growth.

Growth and survival in relation to stocking density in ground plots

The effect of stocking density on the growth of Manila clams at four sites is shown in Fig. 5.7. The clams, initially 0.4 g live weight (13 mm shell length), were kept in net-covered ground frames (0.17 m^2 in area) for 2 years. Size was significantly affected by density only in the River Helford and at Walton-on-the-Naze. At the other two sites, trends of decreasing weight with increase in density were discernible. Differences in growth due to site were greater than those due to density. At three of the sites, market size of 20 g was reached after 2 years, even at densities of up to 1000 clams per m^2.

Survival showed no consistent differences due to density but varied greatly between sites (Fig. 5.8). Survival averaged 60% overall, but ranged from more than 100% (a figure probably due to inaccuracy in estimating initial numbers of spat planted) at Walton, a sheltered stable environment, to 34% in the River Exe, where the soil is mobile and liable to smother unattended clam plots. Crop yield, reflecting both growth and survival at harvest, exceeded 20 kg/m^2 at Walton and Helford in plots seeded at a rate above 1000 clams per m^2 (Fig. 5.9). This trial showed that Manila clams can be re-laid on small plots at densities of up to 1000 clams per m^2 and grown to market size of 20 g within two to three growing seasons. The small loss in growth of indi-

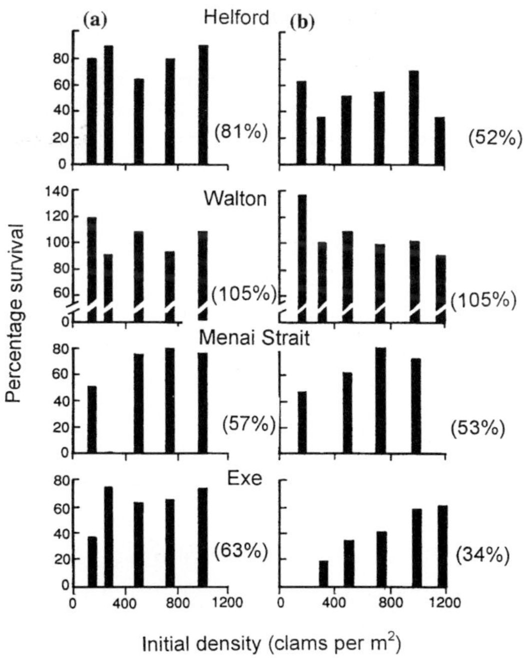

Fig. 5.8 Survival of Manila clams after (a) 1 year and (b) 2 years in relation to stocking density. Mean values are shown in parenthesis.

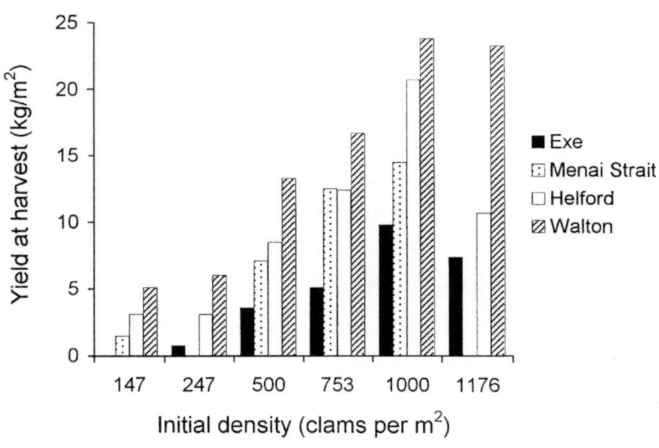

Fig. 5.9 Crop yield of Manila clams in relation to stocking density.

vidual clams is probably compensated for by the cost benefits of harvesting higher density plots. Yields of up to 23 kg/m^2 were recorded, but the fate of clams held at these high densities for a further season's growing is not known. It is apparent that, at sites liable to siltation, regular maintenance of net covers is essential to obtain the best yields.

Preparation of commercial size ground plots

Small bivalves such as mussels, oysters and clams are the preferred foods of shore crabs (*Carcinus maenas*), one of the most common predators in the estuaries and coastal waters of the UK. If Manila clams up to 30–35 mm in size were placed on the sea bed without any form of protection, they would be eaten by crabs within a few tides. Protecting clams is, therefore, an essential part of shellfish farming. The usual way of excluding crabs is to cover the clam beds with plastic netting. This must retain the clams yet allow a good exchange of water containing food to ensure good growing conditions. The net must be sufficiently rigid to prevent the crabs manipulating it and crushing the clams through the net apertures. The importance of this is shown by the amount of protection given to 10–20 mm Manila clams set out in the Menai Strait in summer beneath 5 mm meshes of various types (see Table 10.2). Rigid meshes containing more than 500 g of plastic per m^2 of net are effective crab barriers, but lightweight nets, although providing some protection, are less successful in preventing crab attacks. The effectiveness of lightweight nets can be improved, however, either by creating a gap by raising the net about 5 cm above the soil or by using two layers of net.

Rolls of netting, usually about 2 m wide by several hundred metres long, are available in a range of mesh sizes (5 mm and 10 mm mesh are suitable depending on clam size). A net 2 m wide may be used to cover a plot 1.5 m in width, thus allowing the edges to be buried for good anchorage and preventing entry of crabs or escape of clams. A plot seeded with 100 000 clams at a rate of 500 clams per m^2 therefore requires an area 1.5 m wide by 134 m long.

As a guide, the steps required for establishing a clam plot are as follows:

- Remove large stones, weeds, mussels, crabs, etc., from the surface of the plot.
- Temporarily mark the outline of the plot with posts and string. On soft ground, avoid treading within the boundary lines (Fig. 5.10).
- Dig a trench along the boundary lines. In soft ground, a spade pressed into the soil to blade depth and moved firmly back and forth to create a 'V' shaped groove, 10–15 cm wide at the top, is adequate to bury the net edges to a depth of 20–25 cm (Fig. 5.11).
- Distribute the clams evenly. Seven hundred and fifty clams, counted into a small container, provide a rough visual guide of the quantity to sow on each metre's length of soil.
- Cover the clams with netting.
- Bury the edges of the netting in grooves and backfill (Fig. 5.11).
- Pin the mesh to the soil using wire staples at 1–2 m intervals along the

Fig. 5.10 Marking a plot in preparation for re-laying clams beneath netting.

Fig. 5.11 Burying the edges of the net.

Fig. 5.12 Established net-covered plots of clams. (Photo: D. B. Edwards)

perimeter. Staples with arms of about 20 cm long may be made from 3 mm galvanised wire.

Since the plots (Fig. 5.12) are likely to last for 2–3 years, it will be necessary to exchange the netting for one with a larger mesh to improve the water flow over the clams as they grow. In some circumstances, the netting may be covered with sediment. It is important to keep the deposits to a minimum, especially where small meshes are used. A long-handled squeegee, 50 cm wide, fitted with a semi-rigid rubber strip, is a useful tool for clearing sediment from lays without damaging the netting. Weed growth on the netting, although not hazardous, may reduce water exchange or encourage mud deposition and should be brushed off periodically.

Mechanisation of clam cultivation

Manual net laying or harvesting by digging or raking clams from the soil is labour intensive and therefore suitable only for small-scale cultivation. Large-scale operations need to be machine aided.

In France some clam sites on firm sandy ground permit safe working of tractor-driven machinery. Tractors fitted with rotating brushes enable strips of netting to be cleaned in one operation (Fig. 5.13). Also, machinery has

Fig. 5.13 Brushing clam plots free from weed and sediment in France.

been developed there to facilitate the planting and covering of clams with nets and to harvest clams at market size (Fig. 5.14). Harvesting is usually achieved by the insertion of an angled blade *c.* 10 cm into the soil and lifting the clams and sediment onto a conveyor riddle. This comprises a series of horizontal parallel bars spaced at a set distance apart, which separates the clams of a required size and returns the smaller ones and the soil to the clam plot.

In the UK, harvesting technology has taken a different course with the use of a simple suction device. This is hand-held and used at low tide when the clam beds are covered by water between *c.* 0.2–1.5 m in depth (Fig. 5.15). The suction harvester is a PVC pipe (1.5 m long × 15 cm diameter), open at both ends and with a water inlet inserted near the bottom. The water, supplied via a boat-mounted petrol pump, enters the pipe at a rate of up to 1 m^3 per minute. This creates suction at the mouth of the pipe, which lifts sediment and clams upwards. A mesh bag fitted at the top collects the clams as they emerge from the pipe.

Environmental effects of clam cultivation

In the UK nature conservationists have shown concern that the two non-native species introduced for commercial cultivation, the Manila clam

Fig. 5.14 French harvester.

and Pacific oyster, might pose a threat to the natural ecology of British waters. Their concern lay with the prospect of these bivalves spawning and establishing a self-sustaining broodstock, or competing with natural communities for space and food. Although the Pacific oyster has been cultivated in British waters since the mid-1960s, and natural spat seen on occasions (Spencer *et al.*, 1994), it has not become established. Natural spatfalls of the Manila clam have been recorded in Poole Harbour following exceptionally warm summers in the early 1990s.

As a consequence, an experimental study was set up in the River Exe to answer some of the questions relating to whether clam cultivation has an effect on natural intertidal communities and whether these changes were reversed after harvesting. This was achieved by regularly monitoring the infauna and sediment composition of plots (10 × 1.5 m in area). Samples of the sediment were collected during the 3-year cultivation cycle before harvesting, and for 1 year after harvesting, using the suction device described above. Nine plots were established, three with clams (seeded at 500 per m^2) covered with netting, three with netting but no clams and three control plots without clams or netting. The netted plots received monthly husbandry checks that included the removal of accumulated sediment or weed (the green alga, *Enteromorpha* spp., grew on the netting during the summer).

Fig. 5.15 Harvesting by hand-held suction device. (Photo: D. B. Edwards)

Changes during cultivation

The presence of the netting and the alga growing on it, irrespective of whether clams were present, caused an increase in sedimentation rate compared with the control plots. Thus the netted plots became slightly humped in profile with their central areas elevated about 10 cm above the surrounding levels. Associated with this, the sediment composition of the netted plots altered slightly, showing a small increase in the proportion of silt. A small increase in the organic content and chlorophyll breakdown products of the sediment was probably caused by the presence of the weed and the extra numbers of periwnkles (*Littorina littorea*) grazing on it, and by the extra abundance of deposit-feeding worms beneath the nets.

The numbers of some worm species increased substantially beneath the netted plots irrespective of whether clams were present. This increase occurred within 6 months of net placement and persisted until the clams were harvested, 2½ years after seeding (Spencer *et al.*, 1996, 1997).

Changes caused by suction harvesting

Harvesting created an increased sediment load in the water (4–5 g/l dry weight of sediment within one metre of the harvester) that dispersed to near background levels within 40 m of the device. Some of the particles settled on the sea bed as a thin veneer a few metres wide downstream of the harvester. The immediate effect of harvesting caused a reduction of 80–90% in species diversity and abundance (Table 5.3, and Spencer *et al.*, 1998).

Suction harvesting also left a trench 10 cm deep which took about 3–4 months to fill to background levels. By this time the particle size composition and pigment content of the sediment were similar between plots, indicating a return to the homogeneous nature of the site seen before cultivation began. The animal community recovered slowly during the winter, but by mid-summer natural recruitment had returned species diversity and abundance to levels similar to that of surrounding areas.

Mechanical harvesting of intertidal clams inevitably creates physical disturbance of the sea bed and its associated fauna. In general, studies have shown that mechanical harvesting, including suction dredging, causes large short-term changes to the intertidal habitat. The rate at which recolonisation and sediment structure are restored varies with local hydrography, exposure to natural physical disturbance and sediment stability. In the UK the cultivation of clams in prepared plots is a small industry and poses no real ecological threat. The demand for cockles is much greater. In England and Wales the annual landings in 1990–1994 of managed fisheries averaged 26 000 t, mostly harvested intertidally by suction dredge. It has been estimated that in some areas 150 t of sediment per t of cockles harvested are removed by suction dredging at stock levels of *c.* 50 cockles per m^2 (Rees, 1996). Since cockle fisheries occur largely in exposed areas where the

Table 5.3 Changes to the commonest species of the animal community in the sediment caused by suction harvesting. Values for control plots are shown in parenthesis. Harvesting was done in November 1994

	Species per sample	Animals per m^2
Before suction harvesting	7.9 (5.5)	1812 (760)
After suction harvesting	1.4 (5.5)	238 (760)
8 months after suction harvesting	9.5 (10.1)	3176 (2603)
12 months after suction harvesting	15.6 (16)	482 (577)

substrate and infaunal communities are frequently disturbed by high wave and tidal energy, suction harvesting is unlikely to pose any long-term effect on the ecology of the area. Environmental concern has, however, been directed mainly at sensitive areas where relatively minor cockle fisheries impinge on bird or eel-grass communities.

The management implications of clam cultivation for nature conservation purposes are likely to be varied. Prior consideration of the likely effect of cultivation on the environment can help to resolve some of the problems before they occur. The shellfish cultivator may need to consider rotational use of the clam beds, to limit harvesting to early winter to allow optimum spring larval recruitment or to summer, where recruitment is largely achieved through passive or active movement of juvenile of adult infaunal species (Hall & Harding, 1997).

References and bibliography

Anderson, G. J., Miller, M. B. & Chew, K. K. (1982) *A Guide to Manila Clam Aquaculture in Puget Sound.* Washington Sea Grant Program, Technical Report WSG 82–4.

Anon. (1988) *La palourde – Dossier d'Élevage.* IFREMER, Paris.

Bourne, N. (1989) Clam fisheries and culture in Canada. In: *Clam Mariculture in North America* (eds J. J. Manzi & M. Castagna), pp. 357–82. Elsevier, Amsterdam.

Chew, K. (1989) Manila clam biology and fisheries development in western North America. In: *Clam Mariculture in North America* (eds J. J. Manzi & M. Castagna), pp. 243–62. Elsevier, Amsterdam.

FAO (1999) *Yearbook of Fishery Statistics for 1997.* Vol. **84**. Food and Agriculture Organisation of the United Nations, Rome, Italy.

Goulletquer, P. & Héral, M. (1997) Marine molluscan production trends in France: from fisheries to aquaculture. In: *The History, Present Condition, and Future of the Molluscan Fisheries of North and Central America and Europe, Vol.* **3**, *Europe* (eds C. L. MacKenzie, Jr, V. G. Burrell, Jr, A. Rosenfield & W. L. Hobart), pp. 137–64. US Department of Commerce, NOAA Technical Report **129**.

Hall, S. J. & Harding, M. J. C. (1997) Physical disturbance and marine benthic communities: the effects of mechanical harvesting of cockles on non-target benthic fauna. *Journal of Applied Ecology*, **34**, 497–518.

Hidu, H. & Newell, C. R. (1989) Culture and biology of the soft-shelled clam, *Mya arenaria.* In: *Clam Mariculture in North America* (eds J. J. Manzi & M. Castagna), pp. 277–92. Elsevier, Amsterdam.

Kraeuter, J. N. & Castagna, M. (1989) Factors affecting the growth and survival of clam seed planted in the natural environment. In: *Clam Mariculture in North America* (eds J. J. Manzi & M. Castagna), pp. 149–65. Elsevier, Amsterdam.

Loosanoff, V. L. & Davis, H. C. (1964) Rearing of bivalve molluscs. In: *Advances in Marine Biology*, Vol. **1**. (ed. F. S. Russell). Academic Press, New York.

Mattei, N. & Pellizzato, M. (1997) Mollusc fisheries and aquaculture in Italy. In: *The History, Present Condition, and Future of the Molluscan Fisheries of North and Central America and Europe*, Vol. **3**, Europe (eds C. L. MacKenzie, Jr, V. G. Burrell, Jr, A. Rosenfield & W. L. Hobart), pp. 201–17. US Department of Commerce, NOAA Technical Report **129**.

Newell, C. R. (1991) The soft-shell clam *Mya arenaria* (Linnaeus) in North America. In: *Estuarine and Marine Mollusc Culture* (ed. W. Menzel), pp. 1–10. CRC Press, Inc., Boston.

Nie, Q-Z (1991) The culture of marine bivalve molluscs in China. In: *Estuarine and Marine Bivalve Mollusc Culture* (ed. W. Menzel,), pp. 261–76. CRC Press, Boston.

Rees, E. I. S. (1996) *Environmental Effects of Mechanised Cockle Fisheries: a Review of Research Data.* A report commissioned by the Marine Environment Protection Division, Ministry of Agriculture, Fisheries and Food, London.

Spencer, B. E., Edwards, D. B. & Millican, P. F. (1991) *Cultivation of Manila Clams.* Laboratory Leaflet, MAFF Directorate of Fisheries Research, Lowestoft **65**.

Spencer, B. E., Edwards, D. B., Kaiser, M. J. & Richardson, C. A. (1994) Spatfalls of the non-native Pacific oyster (*Crassostrea gigas*) in British waters. *Aquatic Conservation: Marine & Freshwater Ecosystems*, **4**, 203–17.

Spencer, B. E., Kaiser, M. J. & Edwards, D. B. (1996) The effect of Manila clam cultivation on an intertidal benthic community: the early cultivation phase. *Aquaculture Research*, **27**, 261–76.

Spencer, B. E., Kaiser, M. J. & Edwards, D. B. (1997) Ecological effects of intertidal Manila clam cultivation: observations at the end of the cultivation phase. *Journal of Applied Ecology*, **34**, 444–52.

Spencer, B. E., Kaiser, M. J. & Edwards, D. B. (1998) Intertidal clam harvesting: benthic community change and recovery. *Aquaculture Research*, **29**, 429–37.

Utting, S. D. & Spencer, B. E. (1992) Introductions of marine bivalve molluscs into the United Kingdom for commercial culture – case histories. *ICES Marine Science Symposium*, **194**, 84–91.

6 Oyster cultivation

Introduction

World-wide, oyster landings amounted in 1997 to 3.3 million tonnes, the largest contributor to the total mollusc production of 15.9 million tonnes (FAO, 1999). The 'cupped' oysters (*Crassostrea/Saccostrea* species) form the bulk of world landings, with the Japanese/Pacific oyster (*Crassostrea gigas*) the main contributor to the landings (Table 6.1).

The Pacific oyster is a species of the western Pacific, with China, Japan and the Republic of Korea providing 92% of the landings of nearly 3 million tonnes of this species in 1997. The range of the Pacific oyster has been extended substantially across continental boundaries with intentional and sometimes unintentional movements by man. In the early part of the twentieth century (1912 to 1961), small and large consignments of oysters were moved from Japan to the west coast of North America (Canada and USA), where fisheries quickly developed with the establishment of naturally breeding populations. France also imported substantial quantities of Japanese seed in the late 1960s and early 1970s to supplant it's ailing Portuguese oyster industry, which was suffering from the severe effects of disease. The Pacific oyster quickly became established, breeding naturally in the warmer Atlantic waters of the French coast up to the south Brittany peninsula. The French oyster industry harnesses enormous quantities of spat each year by the deployment of collectors. The spat are grown mainly in containers on trestles and now support massive annual landings of *c.* 140 000 t.

Introductions of the Pacific oyster into the Netherlands, New Zealand and Australia have seen the development of small fisheries with the naturalisation of the species. In New South Wales, Australia, the species is an unwelcome alien since it is competing with the prized native Sydney rock oyster in a number of the major fisheries and is proving impossible to eradicate. Within a decade of its accidental introduction into New Zealand in the 1960s, the Pacific oyster has replaced the native rock oyster (*Saccostrea glomerata*) as the principal cultivated species because of its breeding success under natural conditions.

The Pacific oyster was introduced into the UK from the west coast of Canada and the USA in 1965, 1972 and 1978. Small numbers of brood-

Table 6.1 Main areas of production of cupped oysters (*Crassostrea/Saccostrea* spp.) (FAO, 1999)

Species	Common name	Country	1987	1997	10 year average*
			Thousands of tonnes		
Crassostrea gigas	Pacific/ Japanese oyster	China	400.5	2328.6	1059.8 (220%)
		Korea	303.2	218.1	247.1 (88%)
		Japan	258.8	218.0	242.8 (90%)
		France	135.8	147.2	139.0 (106%)
		USA	42.0	33.2	37.7 (88%)
		World total	1168.2	2994.0	1767.2 (169%)
C. virginica	American or Eastern oyster	Canada	4.7	3.7	4.3 (86%)
		USA	181.0	171.8	149.3 (119%)
		Mexico	50.7	38.5	34.3 (112%)
		World total	236.4	220.3	193.0 (114%)
C. rhizophorae	Mangrove oyster	Caribbean/ C. America	2.2	4.8	4.0 (118%)
C. angulata	Portuguese oyster	Portugal	<0.01	0.6	0.4 (163%)
Saccostrea commercialis	Sydney rock oyster	Australia	7.4	5.1	6.2 (83%)
	World total (all species)		1459.6	3279.4	2024.3 (162%)

* 1997 values as % of 10 year average (1987–1996) are shown in parenthesis

stock oysters (150) were deposited in the quarantine facilities at the Fisheries Laboratory, Conwy, to avoid the risk of introducing pests and diseases (see Chapter 10). Spat derived from the initial broodstock, used for experimental plantings around the coast of England and Wales, confirmed the opinion that the species was hardy and fast growing. On this basis and the belief that the species would not be self-sustaining in the wild, broodstock were given to the commercial hatcheries to broaden the base of the species available for cultivation. The Pacific oysters grown in England, Wales, Scotland and initially Ireland are derived from these importations of broodstock.

The *Crassostrea/Saccostrea* oyster species are genetically quite similar to each other. They all have ten pairs of chromosomes and experimental cross-fertilisations between many of them show egg cleavage and larval development. In some cases, viable offspring have been produced, themselves capable of sexual reproduction. Gel electrophoresis studies have shown that the Australian Sydney rock oyster and the New Zealand rock oyster are suffi-

Table 6.2 Main areas of production of flat oysters (FAO, 1999)

Species	Common name	Country	1987	1997	10 year average*
			Thousands of tonnes		
Tiostrea chilensis	Chilean flat oyster	Chile	0.5	0.3	0.4
Ostrea edulis	European flat oyster	France	2.6	2.5#	1.9
		Ireland	0.6	1.1#	0.8
		Spain	3.2	3.5	3.0
		UK	<0.1	0.6#	0.6
		Greece	0.9	0.4	2.1
		Italy	5.00	0	1.0
		Turkey	<0.1	1.5	1.6
		Netherlands	0.7	<0.10#	0.6
Ostreola conchaphila (*Ostrea lurida*)	Olympia flat oyster	USA	0.03	0.04	0.05
Tiostrea lutaria (*T. chilensis*)	New Zealand dredge oyster	New Zealand	4.8	2.3#	3.3
		World total	23.0	14.5	18.7 (77%)

* 1997 world total values expressed as % of 10 year average (1987–96). # production affected by *Bonamia* disease from early 1980s

ciently similar to be considered geographical races of the same species. Also, surprisingly because of the large distance that separates their natural range, the Pacific oyster and the Portuguese oyster are considered by some scientists to be geographical isolates of the same species. In this case, it is assumed that the Pacific oyster made the journey from Japan to Portugal via merchant ships that frequently plied this route in the sixteenth and seventeenth centuries.

Despite the world-wide meagre landings of flat oysters (Table 6.2), the genus is highly prized gastronomically and commands high prices, currently because of its scarcity value. In recent years, production has been severely curtailed by bonamiasis, a debilitating disease caused by a microscopic organism (*Bonamia ostreae*). This organism invades the blood cells of the European flat oyster and the New Zealand dredge oyster and has caused high mortalities in some fisheries of France, the Netherlands, England, Ireland and New Zealand.

The European oyster industry, traditionally and especially in the nineteenth century, is very mobile, depending on the transfer of adults and juveniles at times of depletion or glut to satisfy market demand. The European flat oyster has been substantially transferred around the coast of the UK and to the Netherlands, France, Norway, Denmark, Germany and Spain. The transfers, which might take place in both directions, are usually from areas of natural abundance and poor growth to ones of low abundance and good growth and have occurred with such regularity that any geographical genetic differences must have been lost from the major fisheries. The American oyster was also regularly deposited in some oyster fisheries in the UK from the early 1870s until 1939 and the Portuguese oyster from 1926 to the 1960s. The practice of importing and planting half-grown seed of these oysters onto good growing and fattening grounds for one season helped the oyster industry to sustain itself at a time of dwindling stocks of flat oysters (see Fig. 1.9 in Chapter 1). Occasional spatfalls of the American and Portuguese oysters occurred in the River Blackwater, Essex, but were not self-sustaining and died out when importations ceased.

A more recent evaluation of the potential of the American oyster in England and Wales was undertaken at a time when the Pacific oyster was under great stress in many areas owing to the water-borne effects of tributyl tin (TBT), a component of many marine antifouling paints. The reintroduction of the American oyster was made in 1984 from a disease-free stock from Chesapeake Bay, USA, via the hatchery of the Fisheries Laboratory, Conwy. In 1986, juveniles of the American and Pacific oysters were planted in trays at six sites and growth and survival was monitored over one growing season. The Pacific oyster out-performed the American oyster at all sites except at West Mersea, where the effects of TBT were especially bad (Fig. 6.1).

Although broodstock of the American oyster was given to the commercial hatcheries, there has been no demand for the species in the UK, especially as the culture of the Pacific oyster has gained momentum in the wake of the ban on use of TBT-based antifouling paints on small boats.

The European flat oyster is the best known and most documented flat oyster species. Its current low abundance throughout European waters reflects the cumulative effects of overfishing, disease, pollution, climate and changed sociological patterns of lifestyle. In the middle of the nineteenth century the European flat oyster was commonplace in the coastal waters of Europe, ranging from the cool waters of Norway to the warmer waters of Spain and Portugal and into the Mediterranean. England and Wales were especially well endowed with oyster beds that stretched in an almost unbroken belt around the coastline and into Scotland. Oysters were also abundant offshore on deep water beds in the North Sea and the Channel. A piscatorial atlas of 1883 (Olsen, 1883) shows a dense bed of oysters 200 miles (320 km) long and up to 70 miles (112 km) wide in the North Sea, stretching from Heligoland in the east to the Dogger Bank in the west. Oyster consumption at this time was commonplace throughout Europe and dedi-

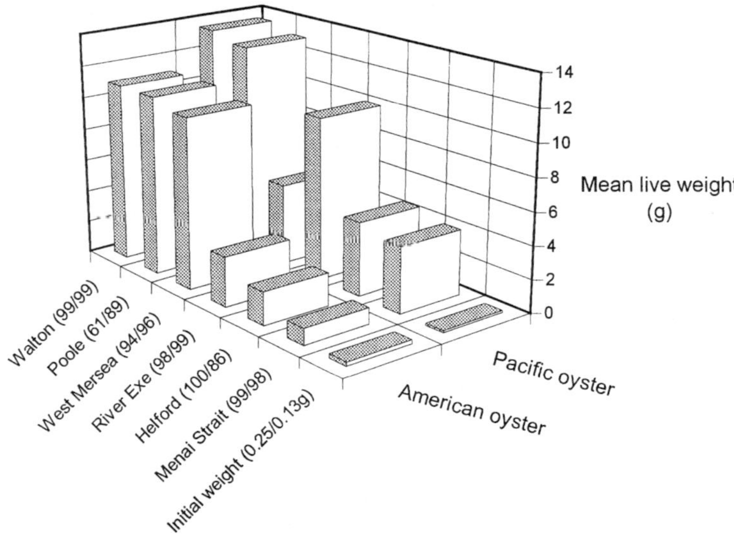

Fig. 6.1 One season's growth of Pacific and American oysters in trays in England and Wales in 1986 (percentage survival is shown in parenthesis against the sites).

cated fishing communities arose to foster the cultivation and fishing of the oysters. Great natural beds existed in the Thames, Medway, and Blackwater estuaries, in the Solent, Langstone, Chichester, Milford, Tenby and elsewhere. There were also very large private beds supported by strong cultivation practices in the Thames, Blackwater, Colne, Roach, Isle of Wight and Cornwall. Oyster production in Europe peaked towards the middle of the nineteenth century. Landings in Cancale, France show a rise and fall related to fishing effort during the Napoleonic Wars (Table 6.3), a situation which may well have been representative of oyster fishing elsewhere on the Atlantic coast of Europe at that time.

One can only guess at the magnitude of oyster production at its heyday in the nineteenth century. Historical records provide glimpses of the size of the industry. Blake *et al.* (1870) provide a whole range of anecdotal statistics, one of which describes oyster consumption in London in 1868 as reaching 700 million. Assuming the oysters had a modest average weight of *c.* 50 g, this figure converts to around 35 000 t. When multiplied up by the consumption in other cities and European countries, annual production may well have been above 100 000 t and near to that of the French Pacific oyster industry today.

The perpetuity of oyster beds, then as now concerned the replenishment of grounds by natural spatfalls which, influenced by the vagaries of nature,

Table 6.3 Decline of oyster production in Cancale, France in the nineteenth century (from Blake *et al.*, 1870)

Mid-point of decade	Landings (millions of oysters)*	
1805	1.2	Low fishing effort during the
1815	2.5	Napoleonic Wars
1825	20.0	Stocks accumulated during wartime
1835	40.8	period provided bountiful harvests
1845	62.3	
1855	28.3	Start of decline due to overfishing
1865	3.0	

* annual landings averaged for ten year period

produced good years and bad and everything in between. In the second half of the nineteenth century, there were 17 very good and 12 very bad spatfall years in the Essex estuaries. The good years of 1857 and 1859 led to a glut of oysters with consequent loss of value. Towards the end of the century, with spatfalls falling permanently to relatively low levels, prices rose and the one-time cheap oyster became a relatively luxury commodity. With the scarcity but continued demand for oysters, the industry turned to foreign stocks to supplement the lack of home grown oysters. French, Dutch and Belgian oysters which had long been imported for re-laying were sought not only for producing market size oysters but also to contribute to the spatting potential of the stock.

Pond culture

The decline in production of flat oysters in Europe led to attempts to develop methods to stimulate production artificially. Pond culture (see Chapter 1), which had been used successfully in Lake Fusara, Italy, in Roman times 2000 years ago, was rediscovered in France, the Netherlands and England in the 1860s. The method provided some degree of environmental control, since the ponds were built near to high water mark and could be filled with sea water and then isolated for a period of time. This was done in early summer, prior to the natural breeding season of the oysters, which were stocked at low density in the ponds along with spat collecting material such as stones, slate, shell and bundles of twigs. In England, ponds were constructed at Hayling Island, Herne Bay, Reculver, Isle of Wight, Lymmington, River Beaulieu, River Orwell, Fagwell and New Brompton. The ponds, ranging in size from 1 to 10 ha and from 1 to 3 m deep, were variously successful, some phenomenally so, producing many tens of millions of spat in some years. Generally spat production was insufficiently regular to enable the industry to rely on the method for restocking the grounds and the ponds fell into disuse. In modern times, the method has been successfully applied in Cork Harbour (Republic of Ireland). There, 22 butyl-lined ponds, each holding 1000 m^3 of sea water, produce sufficient numbers of spat, collected on mussel shell, to sustain a small fishery with a production of about 300 t of market size oysters per year.

In Norway, a variant of pond culture in fjordic polls was developed towards the end of the nineteenth century. The poll, an offshoot of a fjord, has a narrow entrance with a shallow sill. Two types of poll were used for oyster cultivation, relatively small ones (breed-polls), 1–5 ha in size and 5–10 m deep, for breeding, and larger ones (spat-polls) up to 40 ha in area and 20 m deep for growing the spat to market size. Water flow across the sill of the breeding poll was controlled by a sluice gate and wooden barrier built across the shallow entrance. In its closed position, fresh water accumulated on the surface of the poll, providing a greenhouse effect that warmed the underlying sea water to above 20°C in the summer. Oysters and spat-collecting bundles of birch twigs were suspended in the warmed sea water.

Once collected, the spat were transferred to the spat-polls, where growth was more rapid, enhanced by the regular exchange of water from the fjord. The unpredictability of spatfall in the polls led to a decline in their use, from 25–30 polls in the early 1900s to one or two in the mid-1980s. Commercial spat production from hatcheries has not succeeded in Norway, but an interesting semi-intensive spat production method, using mesocosms, has shown promise (Naas, 1991). These comprised plastic bags (7–12 m^3 in volume), immersed in the fjord, in which sea water and natural food and oyster larvae were enclosed. Temperatures of 20–24°C could be maintained during the summer, and by manipulating nutrient supply, light availability and water stratification, spat production was improved further.

The use of ponds for oyster culture have not only been confined to spat collection. In some areas of France (e.g. Marennes-Oléron Bay) shallow ponds (known as claires), up to 0.5 m deep, are used for fattening oysters prior to marketing (Fig. 6.2). The claires, constructed on salt marshes near to high water mark of spring tides, are made with substantial mud walls and firm bottoms onto which the oysters are laid in baskets for 1–2 months. The ponds are filled with sea water at high tide via a complex of channels connecting them to the main channel. The sea water supply, controlled by a system of sluices and valves, may be stopped during neap tides to encourage the blooming of the phytoplankton species, *Haslea* (*Navicula*) *ostrearia*. The presence of this plant species leads to a greening of the oyster gills, a condition that is highly desirable to the French market, creating high-priced oysters, which are said to be especially tasty. Some of the creeks of the

Fig. 6.2 French claires showing channel used for filling and emptying.

Blackwater and Colne also produced greening of oyster gills, a condition which at the turn of the nineteenth century was not favoured by the English market, which preferred oysters with white flesh. Oysters from these areas were, therefore, sent for re-laying for several months on the Whitstable beds, where they lost their green colour.

A similar system of ponds (oyster pits), now largely derelict, evolved on the salt marshes of the Essex creeks (see Benham, 1993). These were used for storing oysters for market in the autumn and for over-wintering immature oysters for return to the beds in the following spring. Harsh winters, producing freezing ice, low salinities and smothering bottom deposits in the shallow creeks, were sufficiently frequent to justify the extra costs of moving the oysters from the beds to the pits. By allowing frequent water changes at high tides, the oystermen were able to reduce the effects of adverse weather conditions on the oysters.

Natural flat oyster fisheries

Despite the current (1997) depressed state of the major flat oyster fisheries in Europe and New Zealand, caused largely by the disease bonamiasis, the industries continue to operate, albeit at lower production levels and under more stringent management controls.

In England the major fisheries are confined to the River Fal (Cornwall) and the Solent (Hampshire), with derived fisheries in the Essex creeks dependent on re-laying seed oysters from the other two fisheries. The River Fal was especially hard hit by bonamiasis in 1982, with a high infection level requiring a ban on transferring the seed to other areas of the UK. Although *Bonamia* has been found in some Solent stock, the level of infestation is sufficiently low to permit the transfer of seed to other disease areas, e.g. the Essex creeks. However, re-laying stock in these areas was conditional on the application of certain management controls designed to reduce the build-up of disease between seasons. The controls specified that:

- the oysters must be re-laid at low density (density not specified but in France advice was 1–2 oysters per m^2) and re-laid for one season only,
- unsold stock remaining at the end of the season must be removed and destroyed, and
- natural spatfall must be discouraged (i.e. by not re-laying cultch).

These controls were successful and allowed oyster culture to continue.

The Solent fishery has supplied oysters to other areas better suited to growth and fattening since the early days of oyster cultivation in the nineteenth century. Nowadays the major part of the catch is sent to Spain and France and a minor part (<100 t in the mid-1990s) to private layings and a several fishery in Essex for further growing. The Solent fishery is largely divided by parliamentary orders into a regulated fishery and two several fisheries (i.e. the public rights are severed completely). The regulated fishery, 17 195 ha in area, is managed by the Southern Sea Fisheries Committee who issue licences to fish for oysters and impose controls on type and size of dredge, fishing season and minimum landing size of oysters etc. In the period 1981–94, the official landings from this fishery averaged 203 t per year. The several fisheries are managed by groups of local fishermen, acting via elected committees who impose their own rules for catching oysters, including a very restrictive fishing season that lasts for just a few weeks each year. These two fisheries, 485 ha in area, had an average combined annual catch of 142 t for the 20 year and 13 year periods from their inception, to 1994–95. Because of the exclusive nature of the fisheries, the overseeing government department in annual reviews expects to see efficient management practices in operation that are likely to lead to catches above those that might be expected naturally. For example, the deposit of cultch such as mussel, cockle or scallop shell, in the early summer, is an expected and age-old practice to encourage settlement of spat on the fishery.

Other, once highly productive, oyster fisheries in England are showing signs of re-regeneration again. For example, the Blackwater estuary in Essex produced significant spatfalls in the 1990s, causing a resurgence of interest by local oystermen in cleaning the beds and re-laying cultch on the Tollesbury and Mersea several fishery. This fishery, together with many other shallow water fisheries on the east coast, was hard hit by the extremely cold winter of 1962–63, which killed most of the stock, and again in 1982 by the appearance of *Bonamia* in oysters re-laid in the adjacent creeks. The oystermen's efforts to diversify into the cultivation of Pacific oysters was thwarted by the presence of TBT in the water, leached from the antifouling paints applied to the hulls of the large numbers of yachts in the area. TBT had a severe effect on growth, producing golf ball-shaped Pacific oysters and a less obvious but none the less serious effect on flat oysters by reducing their breeding capability. The nation-wide ban on the use of TBT-based anti-fouling paints on small boats in 1987 led quickly to an improvement in water quality. Within a year or two, Pacific oyster growth and flat oyster reproduction returned to normal, a situation that is leading to a healthy revival of oyster cultivation in the area.

Flat oyster culture techniques have not changed very much from the nineteenth century. The most obvious developments have occurred in spat collection. The deposit of shell has been used extensively to the present day. In the latter part of the nineteenth century, limed roofing tiles were brought into use in France, their use spreading to other large oyster-producing countries such as the Netherlands. The tiles were placed in the low intertidal zone of areas known to produce reliable spatfalls in June and July. Spat collected on the tiles were removed by hand in the April of the following year. They were grown intertidally in trays for a further season and then re-laid on shallow culture plots. In 1885, as many as 30 million tiles were placed on the oyster fisheries of the Netherlands. The annual labour and cost of preparing tiles, i.e. cleaning, scraping and dipping them in lime mixture, however, precluded their widespread use and they lost their popularity except in the Arcachon region of France, where they are used to some extent even today (Fig. 6.3). Cockle shell was a cheaper alternative to tiles and could be laid on the subtidal beds to widen the spat-collecting catchment area. In the Netherlands between 1924 and 1930, 40 000–50 000 m^3 of cockle shell were laid each summer. However, a link between using cockle shell, which slowly breaks down on the beds, and the persistence of shell disease in flat oysters caused by a fungus, *Ostracoblabe implexa*, was found. Cockle shell was therefore banned in 1936 and was replaced by mussel shell, which degrades much faster. Today, Dutch oyster beds, confined to areas of the Oosterschelde (1700 ha) and Lake Grevelingen (380 ha), produce a few million flat oysters each year. Spat are collected on special plots where mussel shell is laid at a rate of 30–60 m^3/ha (Dijkema, 1997).

Spat collecting on bundles of twigs of wood is still practised on a small scale in Italy, Norway and Croatia, although new materials are being experi-

Fig. 6.3 Liming pits used for coating tile collectors prior to deployment in Arcachon.

mented with continually. Once collected, the spat are removed from the twigs and grown in plastic mesh containers (Norway) or cemented to pieces of wood or ropes (Croatia) and grown to market size, suspended from a longline, raft or frame.

The New Zealand dredge oyster fishery, located in the Foveaux Strait between Stewart Island and South Island, is a natural fishery regulated by limiting landings, number of vessels, season and minimum size, etc. Although world landings (Table 6.2) indicate that it is the second most important flat oyster species, next to the European flat oyster, the statistics hide the fact that these landings are for experimental rather than commercial use. The sustainable catch limit of the fishery, set in 1975 at 115 000 sacks per year (1 sack = 79 kg = 774 oysters), was maintained at that level until 1986 when *Bonamia ostreae* (serologically different from the European strain) was first identified in the population. Between 1986 and 1990, the oyster population of the Foveaux Strait declined by 58%, owing largely to mortalities caused by *Bonamia*, and the landings were down to *c.* 46 000 sacks per annum (Cranfield *et al.*, 1991). With continuing mortalities in the early 1990s, the beds were closed to commercial fishing in 1993 but with a small allowable catch (1200 sacks in 1995–96) to supply an experimental spat-producing programme.

The cupped oyster fisheries

The cupped oysters, comprising the *Crassostrea* and *Saccostrea* genera, are the most widely cultivated and important commercial oyster species in the world. They are found naturally in the warmer waters of the world, where man has harnessed their prolific reproductive capability to produce enormously successful oyster fisheries. Transfers and introductions of stock, especially that of the Pacific oyster and Sydney rock oyster, have widened their ranges, often leading to the establishment of natural fisheries (e.g. Sydney rock oyster from Australia to New Zealand, later to be supplanted by the Pacific oyster; Pacific oyster from Japan to Canada, USA, France and Australia). The Pacific oyster has also been transplanted to countries with temperate waters (e.g. European countries bordering the Atlantic coast e.g. Norway, Denmark, Germany, UK, Ireland, Spain etc) to produce small but viable industries based on hatchery production of seed. A small Pacific oyster fishery around the German island of Sylt (Wadden Sea) can be perpetuated only by overwintering the oysters ashore in tanks to avoid the ice conditions.

The following section provides an insight into the cultivation practices of some countries, originally unique but becoming more uniform in character with copied technologies.

Japan

Although the origin of Pacific oyster cultivation in Japan can be traced back four centuries, the modern industry took off in the early 1950s. Prior to this expansion and today in local fisheries, relatively simple forms of cultivation were used. These included ground culture and stick culture, both undertaken intertidally in sheltered bays and estuaries. With stick culture, branches of trees were inserted vertically in the ground to catch spat, which were then left to grow in situ to market size or were knocked off onto ground layings.

The basis for present day cultivation began in the 1920s with the concept of 'hanging' culture introduced in Hiroshima Prefecture. This method employed the use of spat-collecting oyster or scallop shells threaded onto strings of wire (rens) with bamboo spacers between each shell (Fig. 6.4). The rens were suspended from a bamboo framework attached to poles inserted into the sea bed near to low water mark. Hanging culture provided considerable advantages over the conventional methods, including improved growth and fattening, three-dimensional use of the water, independence of the type of sea bed, use of deeper water immediately below low tide and

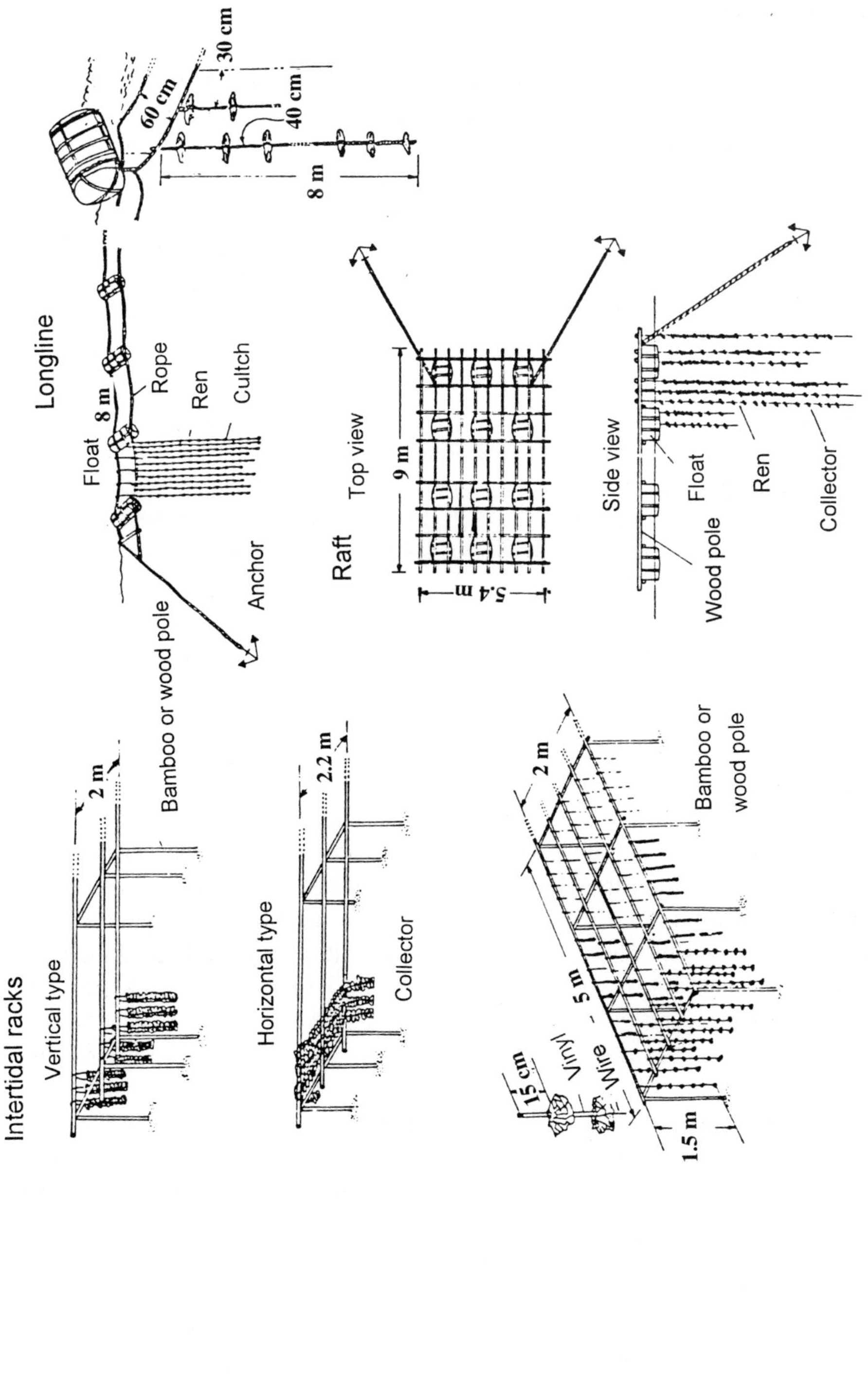

Fig. 6.4 Seed collecting methods used in Japan.

the avoidance of benthic predators. Shortly after settlement, the spat were moved to intertidal racks situated at progressively higher levels of the beach to undergo a period of hardening off. This weeded out the weaker spat and hardened off the survivors, making them more durable to environmental change. After the hardening off period, which usually took place for up to 6 months during the winter, the rens were returned to racks lower on the beach to allow the oysters to grow to market size. This cycle of events is still used today, albeit on a reduced scale. The main departure has been the development of offshore spat collection and on-growing techniques using rafts and longlines, respectively in Hiroshima and Miyagi Prefectures, the two main oyster growing areas. The Japanese oyster industry is largely based on the production of shucked meats. Hence the irregularity of shell shape associated with oysters grown to market size on cultch is of little importance compared with the French industry, for example, which sells most of its oysters in the shell.

Raft culture is centred in the Seto Inland Sea, Hiroshima Bay, an area ideally suited to this method because of the high primary productivity of the water and the calm conditions and shelter from offshore winds provided by the many islands across the bay. A typical raft, measuring 20 m × 10 m, of bamboo poles on supporting beams and plastic floats, might be linked by chain with four others in series and anchored to the sea bed by two or three 10 t concrete blocks at each end. Around 1200 spat-collecting rens, 2 m long bearing 60–70 shells spaced 1–2 cm apart using plastic tubes, are attached to the raft (Fig. 6.4). The timing of deployment of the rens is aided by scientific surveillance to indicate the presence of an appropriate number of eyed larvae in the water or spat on test collectors. Following a period on hardening racks ashore, the shells are re-strung at about 20 cm intervals on a 9 m long wire and returned to the rafts for on-growing to market size. Hiroshima Prefecture is the main oyster growing area with *c.* 70% of the total Japanese production. In 1986, there were 2.5 million m^2 of raft surface area in the Prefecture, producing 154 000 t of oysters (Anon., 1989). Persistently high production, however, led to overstocking problems. The carrying capacity of the area was exceeded and, with reduced growth increasing the cultivation cycle to over 1 year, management strategies have been introduced to control the number of rafts (Héral, 1993).

Longline culture is carried out mainly in the coastal rias and offshore waters of Miyagi Prefecture. This relatively new method, introduced around 1952, quickly overtook rack and raft culture, since it enabled the farmers to exploit the offshore, rougher waters. A longline comprises two parallel ropes buoyed at the water surface by plastic floats of sufficient number and buoyancy to support the load of shells and oysters (see Chapter 7), and anchored at each end with concrete blocks. Strings of shells with attached seed are suspended from the ropes (Fig. 6.4). The exposed conditions in Miyagi Prefecture have led to a modified design of ren from that used in Hiroshima Prefecture. A synthetic string replaced the wire and the shells were secured by wedging them within the strands of the string to avoid wear and slippage downwards. Even so, the rougher conditions caused a greater loss rate from oysters falling off the shells, a situation that was partly alleviated by shortening the on-growing cycle to 1 year. The general procedure for producing market oysters is much the same as that used in Hiroshima Prefecture, with intertidal hardening racks used over winter prior to moving the shells back to the offshore longlines.

Although Miyagi Prefecture is the second largest oyster producer in Japan with about 20% of the total, it produces most (70%) of the country's seed requirements.

China

Despite a recorded antiquity of oyster culture dating back 2000 years, it is only in the last 40 years or so that China has grown to be the world's largest producer with landings of 2.3 million tonnes in 1997 (FAO, 1999). FAO statistics (Table 6.1) list *Crassostrea gigas* as the component species of oyster production in China. That this is not so partly reflects the confusion of nomenclature of oysters in the western Pacific. Many species of oyster occur in China and at least three are grown commercially. Their scientific names, still under debate but following recent opinion (Carriker & Gaffney, 1996), are *Alectryonella plicatula* (= *Ostrea plicata*) the plicate kitten's paw oyster, *Crassostrea ariakensis* (= *C. rivularis*) the Suminoe oyster, and *C. talienwhanensis* (= *C. gigas*).

Oyster cultivation takes place along the whole length of the Chinese coastline. This is about 18 000 km in length, with climatic conditions ranging from temperate in the north to semi-tropical in the central East China Sea area and tropical in the southern South China Sea area. *Alectryonella plicatula* (over 80% of the landings) and *Crassostrea ariakensis* are the most common of the oyster species cultivated. They are warm water species and are cultivated in the Provinces bordering the central and southern parts of the coastline. *C. talienwhanensis* is a cooler water species forming a mainly wild fishery in the north with minimal husbandry involving some bottom preparation and predator control.

Diverse methods of cultivation, largely involving ancient bottom techniques but with an increasing awareness of the advantages of off-bottom culture using racks, rafts and longlines, are used. The ancient methods of rock and stake culture are essentially intertidal methods performed in sheltered, shallow waters of bays and estuaries.

Rock culture

This method, used on hard substrates, has a number of variants ranging from the deployment of boulders and rocks to simple 'stone bridges' that raises some of the slabs off the bottom (Figs 6.5, 6.6; from Cai & Li, 1990). The stones, frequently natural marble flagstones, are cleaned, sometimes coated with limewash, dried and set out immediately before the expected spatfall in the early summer. Up to 60 000 stones per hectare may be used which, depending on species and area, may yield up to 3 t of oyster flesh. Since the 1950s, lime-coated concrete posts and tiles have been introduced for the culture of *C. ariakensis*. Up to 145 000 tiles per hectare, set out 5 cm apart in 12 m rows, with 70 cm between rows may be used. After spatfall, the tiles are transferred to better growing grounds and spaced further apart to ensure good growth. The spacing procedure usually takes place each year of the 4-year cycle to harvesting. Expected production can be up to 15 t of meat per hectare (Cai & Li, 1990).

Stake culture

This method, using bamboo or wooden stakes, is suitable for exploiting soft-bottom areas. The stakes, 1.2 m long × 20–30 mm in diameter, are ini-

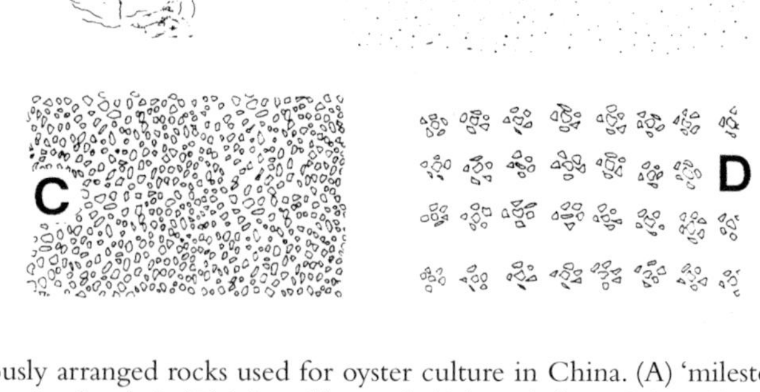

Fig. 6.5 Variously arranged rocks used for oyster culture in China. (A) 'milestones', (B) 'small house', (C) 'stars in the sky', (D) 'plum blossom'. (Adapted from Cai & Li, 1990.)

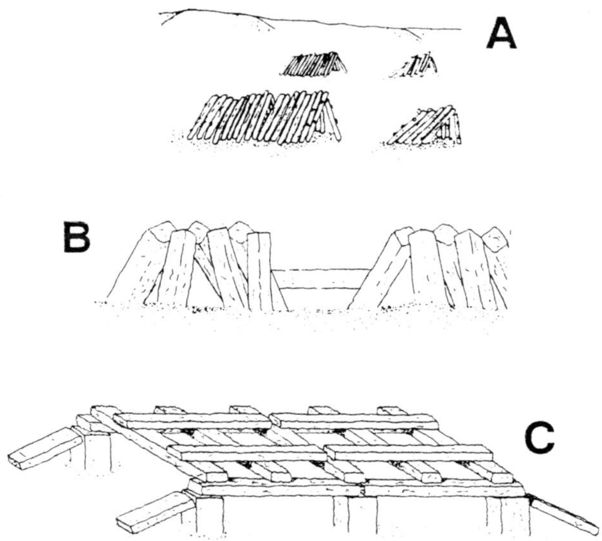

Fig. 6.6 Stone slabs used for oyster culture in the south of China. (A) 'scissors' and (C) 'stone bridge' arrangements are used for spat collection. (B) 'oyster posts' are used for growing oysters. (Adapted from Cai & Li, 1990.)

tially set out in bundles of 4 or 5 in a tent arrangement (Fig. 6.7) for spat-collecting. These are arranged in rows with around 200 000 stakes per hectare (Nie, 1991). After spatfall, the stakes are relocated more thinly, and in a different format (Fig. 6.7), to achieve better growth. The growth cycle can take up to 18 months, depending on time of spatfall, yielding about 6 t of meats per hectare (Nie, 1991) or 0.5 t of meats per 10 000 stakes (Cai & Li, 1990).

Other methods

Although oyster production in China is still achieved largely by the relatively simple technology of rock and stake culture, rafts and racks are likely to play an increasingly larger part in the future. In the two main production areas, 65% of oyster production is from the stone bridge method, 20% from rock culture and 15% from bamboo stakes (Fujian Province); and 90% using concrete culture and 10% from racks and rafts (Guandong Province) (Cai & Li, 1990).

France

The Portuguese oyster became a popular species for cultivation in France following its introduction in 1860 with oysters imported from Portugal to

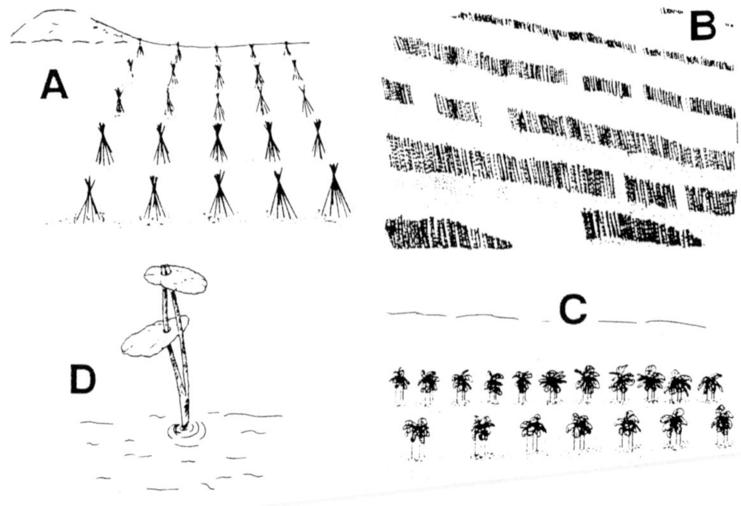

Fig. 6.7 Variously configured bamboo or wooden stakes used for oyster culture in China. (A) 'tent' and (D) stakes with shells, are used for spat collection. (B) 'fence' and 'dash' patterns, and (C) 'Cha Huong' are methods used for growing oysters. (Adapted from Cai & Li, 1990.)

Arcachon Bay. It quickly became established in the Gironde estuary and during the next 20 years spread naturally along the Atlantic coast, eventually reaching a northerly limit in France on the south Brittany coast (Goulletquer & Héral, 1997). Portuguese oyster fisheries in north Brittany and Normandy were established with spat collected from the two main spat-collecting centres in Arcachon and Marennes-Oléron Bays. By the 1920s, the Portuguese oyster had largely replaced the flat oyster on the collectors in Arcachon Bay. Production increased to just over 100 000 t in the mid-1950s with the highly favourable growing areas of Arcachon and Marennes-Oléron Bays becoming heavily overstocked. This led to poor growth, increased mortality and a decline in production. By the mid-1960s an iridiovirus disease spread in the main fisheries, creating massive mortalities and the final disappearance of the Portuguese oyster in the early 1970s.

The socio-economic problems associated with the demise of such an important oyster species to the shellfish industry, led to the introduction of the Pacific oyster from Japan in 1972. This strategy was so successful that production from natural spatfalls reached 50 000 t within a few years and exceeded 100 000 t by the early 1980s, a level that has been maintained to the present time (Table 6.1).

The essence of success of the French oyster industry is the reliability and regularity of the spatfalls in Arcachon and Marennes-Oléron Bays, where up

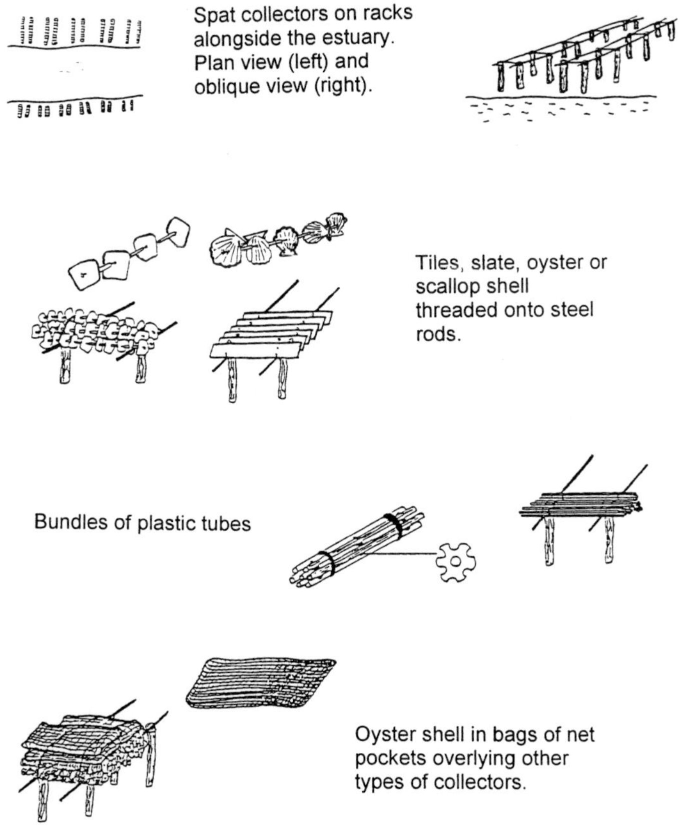

Fig. 6.8 Different types of spat collectors used in France (adapted from Bertomé *et al.*, 1984).

to 25 trillion (25×10^{12}) spat are said to settle each year, and in the Gironde Estuary. Monitoring larval abundance and environmental conditions are key factors in the timing of deployment of spat collectors to optimise spatfall. As with the one-time Portuguese oyster industry, Pacific oyster seed are transferred from the natural production areas to form substantial fisheries in traditional growing areas in the Mediterranean Sea, north Brittany and along the Normandy coast. Spat-collecting methods are numerously varied, showing a progression from traditional stone, slate and shell to limed tiles and, nowadays, plastic (PVC) tubes (Fig. 6.8). The traditional materials, such as slate and oyster and scallop shell, are threaded onto iron rods at up to 100 units per metre or the shell placed in pockets of plastic netting. The PVC tube collector, *c.* 200 cm long by 3 cm diameter, has a corrugated surface to increase surface area and provide a microcosmic sheltered environment for settlement. The tubes are strapped together in bundles of seven. All of these different types of collectors are deployed on trestles, 3 m long and 0.5 m high, near to low water mark. The initial spat collector density is 50–100 per m of trestle, reduced after 1 year to 8–10 per m. When 2–3 cm long, the spat are removed from the collectors and grown in bags (pôches), measuring 1 m long × 0.5 m wide, on trestles in the intertidal zone. Bags-on-trestles is the most commonly used method of Pacific oyster culture in France. Initially, oysters are stocked at 5 kg per bag, reaching 15–20 kg at market size of 50–100 g per oyster. Management regimes include controlling siltation (within the bag and beneath the trestle), stocking density, predators and weed growth on the surface of the bag (see Chapter 4).

The hugely successful Pacific oyster industry in Marennes-Oléron Bay, as in the past with Portuguese oysters, has led to the carrying capacity being exceeded. In the early 1990s, the standing stock amounted to over 120 000 t of cultivated and 22 000 t of wild stock (in Goulletquer & Héral, 1997). Reduced growth, flesh content and higher mortalities of the oysters ensued, leading to a requirement for reduced stocking densities in the area.

Australia

The Sydney rock oyster has traditionally formed the backbone of oyster cultivation in Australia. It is estuarine in habit, occurring naturally on the east coast from Wingan Inlet in eastern Victoria, through New South Wales to Moreton Bay in Queensland. Since a failed attempt to cultivate oysters in Georges River, New South Wales, in 1870 using a system of canals, systematic cultivation has evolved over the years through three distinct methods. With the earliest method, rocks and stones were laid on the foreshore in an upright position or horizontally on supports to collect spat, which were then grown to market size. This cumbersome method was supplanted by intertidal bottom culture, with leases hardened with shell and other materials to prevent the oysters from sinking into the mud. The stock was tended by hand and harvested by hand or by dredge. A small number of such leases are still in operation in New South Wales.

Within the last 50 years, the oyster industry has been dependent on stick and tray culture. This is essentially an intertidal method with the level of cultivation chosen as a compromise between growth and survival of the oysters and infection of the shell by the mud worm (*Polydora websteri*). Initially mangrove sticks tied in bundles and placed intertidally to catch spat were used. These were then laid on racks to grow to market size. Hardwood sticks (turpentine dipped in tar to resist marine borers) eventually replaced the mangrove sticks. The sticks, 180 cm long × 2.5 cm square, were nailed 10 cm apart and 5 or 6 layers deep onto a frame to form a 'crate' (Fig. 6.9).

Fig. 6.9 Tar-coated turpentine sticks newly placed on intertidal racks in readiness for spat collection. Port Stephens, New South Wales, Australia.

The crates were placed on intertidal racks in spat-catching areas shortly before the spat settlement period (January–May in New South Wales). After settlement, the crates were transferred intact to intertidal racks in areas suitable for growing. The intact crates protect the small spat from predation by fish. When the spat reach 2–3 cm and are sufficiently large and robust to withstand predation by fish, the sticks are laid out individually, 20 cm apart on the racks (Fig. 6.10) for the oysters to grow to market size, usually 1–2 years later.

At harvest, the oysters are removed from the sticks and graded into sizes suitable for the plate (about 20 oysters per kg) or for bottling (about 30 oysters per kg), with the small ones returned to the racks for further growing. On some farms the oysters are removed from the sticks at a smaller size and grown in trays on racks for 3–15 months to produce marketable individuals with a better shape and flesh content. Growing sites were frequently surrounded by a protective barrier of heavy poles driven into the sea bed supporting netting to exclude predatory fish (rays), and wood boards at the water surface to reduce wind and wave action.

A reassessment of the use of hardwood sticks, with an annual deployment of 3 million in the mid-1980s, and the high labour and costs involved in tarring them each year, has led the industry to experiment with plastic collectors including the French PVC tube.

Fig. 6.10 Individual sticks bearing oyster spat. Port Stephens, New South Wales, Australia.

The industry produced around 150 million Sydney rock oysters and 25 million Pacific oysters in the mid-1980s. Sydney rock oyster production is centred on New South Wales and southern Queensland. Many estuaries and inlets of this coast produce Sydney rock oysters, but the main centres are Wallis Lake, Port Stephens (the main spat-producing area), Brisbane Water and the Hawkesbury and Georges Rivers, which together produce around 70% of the total.

The farming of the Pacific oyster began in Tasmania in the 1970s and it is now cultivated in South Australia, Victoria and New South Wales. Cultivation in most of these States is dependent mainly on hatchery seed, but since 1984–85 the Pacific oyster has become well established in Port Stephens with such significant spatfalls that the industry became seriously alarmed at the possibility of the species ousting the native Sydney rock oyster. Although the NSW industry has tried unsuccessfully to eradicate the Pacific oyster, it is learning to live with this unwelcome alien as it contributes progressively more to the production statistics.

North America

A number of oyster species form commercial fisheries on this continent. On the west coast, the native Olympia flat oyster supports a small and diminishing fishery (Table 6.2) in Puget Sound, Washington State, while the introduced Pacific oyster has become an important and well established fishery in the Georgia Strait, British Columbia and in Puget Sound (Table 6.1). The most important commercial species, however, the American or eastern oyster, occurs on the east coast of North America. Its distribution in the western Atlantic ranges from the Gulf of St Lawrence, Canada to the Gulf of Mexico, the Caribbean, Brazil and Argentina.

Cultivation of the American oyster occurs in estuaries and coastal areas of low salinity, mainly in the USA, where at one time it formed the world's major oyster fishery. The distribution of production in 1989 was: Canadian Maritimes 1%, USA Atlantic States (Maine to Georgia) 32% and Gulf States (Florida to Texas) 67% (from MacKenzie, 1996). Cultivation is based mainly on fisheries derived from natural spatfalls with hatchery seed contributing less than 1% of total production.

On the Atlantic and Gulf coasts of the USA, the American oyster is fished from public grounds rather than farmed on privately leased beds. Traditional culture involves the placement of shell cultch (oyster shells from the shucking industry or dredged fossil shells) in good settlement areas. Once the seed has set on cultch, a variety of strategies are used to grow it to market size. In some areas, the seed is left on the grounds untouched until the oysters are at marketable size of around 75 mm. In other areas the seed, when large enough (2–4 cm), are transplanted to growing beds, remaining there for $1\frac{1}{2}$–5 years until market size. Transplanting the oysters by dredging breaks up the clusters of seed into smaller numbers, which improves survival, shape and saleability of the final product. It also enables the operator to re-lay at lower densities, usually within the range of 45–70 m^3/ha (MacKenzie, 1996). The quantities of cultch re-laid each year vary between fisheries. In Louisiana, the largest producer of oysters (5.2 million kg of meats in 1989) in the USA, nearly 800 000 m^3 of shells were laid on the public grounds in the early 1990s. There does not appear to be a standard recipe for re-laying quantities, but information from fisheries in Connecticut and Florida indicates that shell is re-laid at a rate of 80–250 m^3/ha.

Despite a long history of introductions into other countries, the American oyster is only known to have become established in Hawaii (Pearl Harbour, Oahu) and in British Columbia (Nicomekl River) outside of its normal range. Cultivation of the American oyster in the USA has been diminishing for nearly a century, with its initial decline probably a result of overfishing and ineffective management. Since 1960, the rate of decline has been greatest on the Atlantic coast, from 7.5 million to 1.5 million bushels (264 bushels = 1 m^3), and with little or no change on the Gulf coast (4–3.5 million bushels). The decline on the Atlantic coast has been due to the diseases, *Haplosporidium nelsoni* (MSX) and *Perkinsus marinus* (dermo), which have devastated stocks in the major fisheries of Chesapeake Bay and Delaware Bay.

References and bibliography

Anon. (1989) Making the Best Possible Use of the Oyster's Propagative Power. *Fisheries Journal*, **28**. Yamaha, Tokyo, Japan.

Benham, H. (1993) *Essex Gold, the Fortunes of the Essex Oysterman*. Essex Record Office.

Berthomé, J.-P., Prou, J., Razet, D. & Garnier, J. (1984) First approach of provisional estimation of the potential production of the cupped oyster, *Crassostrea gigas*. *Haliotis*, **14**, 39–48.

Blake, J. A., Francis, F., Hart, G. W. & Brady, T. F. (1870) *Report of the Commission Appointed to Inquire into the Methods of Oyster Culture in England and France*. HMSO, Dublin.

Cai, Y. & Li, X. (1990) Oyster culture in the People's Republic of China. *World Aquaculture*, **21**, 467–72.

Carriker, M. R. & Gaffney, P. M. (1996) A catalogue of selected species of living oysters (Ostreacea) of the World. In: *The Eastern Oyster*, Crassostrea virginica (eds V. S. Kennedy, R. I. E. Newell & A. F. Eble), pp. 1–18. A Maryland Sea Grant Book, Maryland.

Cranfield, H. J., Doonan, I. J. & Michael, K. P. (1991) *Assessment of the Effects of Mortality due to* Bonamia *on the Oyster Population in the Foveaux Strait in 1990 and the Outlook for Management in 1991*. New Zealand Fisheries Assessment Research Document **91/18**.

Djikema, R. (1997) Molluscan fisheries and culture in the Netherlands. In: *The History, Present Condition, and Future of the Molluscan Fisheries of North and Central America and Europe*, Vol. **3**, Europe (eds C. L. MacKenzie, Jr, V. G. Burrell, Jr, A. Rosenfield & W. L. Hobart), pp. 115–35. US Department of Commerce, NOAA Technical Report **129**.

FAO (1999) *Yearbook of Fishery Statistics for 1997*. Vol. **84**. Food and Agriculture Organisation of the United Nations, Rome, Italy.

Goulletquer, P. & Héral, M. (1997) Marine molluscan production trends in France: from fisheries to aquaculture. In: *The History, Present Condition, and Future of the Molluscan Fisheries of North and Central America and Europe*, Vol. **3**, Europe (eds C. L. MacKenzie, Jr, V. G. Burrell, Jr, A. Rosenfield & W. L. Hobart), pp. 137–64. US Department of Commerce, NOAA Technical Report **129**.

Héral, M. (1993) Evolution of oyster aquaculture: problems and perspectives. ICES CM 1993/F: 46.

MacKenzie, C. L. Jr (1996) Management of natural populations. In: *The Eastern Oyster*, Crassostrea virginica (eds V. S. Kennedy, R. I. E. Newell & A. F. Eble), pp. 707–21. A Maryland Sea Grant Book, Maryland.

Naas, K. E. (1991) A semi-intensive method for spat production of the European flat oyster (*Ostrea edulis* L). *Aquaculture Engineering*, **9**, 447–51.

Nie, Q.-Z. (1991) The culture of marine bivalve molluscs in China. In: *Estuarine and Marine Bivalve Mollusc Culture* (ed. W. Menzel), pp. 261–76. CRC Press, Boston.

Olsen, O. T. (1883) *The Piscatorial Atlas of the North Sea, English and St George's Channels*. Taylor and Francis, London.

7 Mussel cultivation

Introduction

World mussel production reached 1.36 million tonnes in 1997 (Table 1.1), constituting 9% of the 15.9 million tonnes total mollusc landings (FAO, 1999). The top five producing countries, China, Spain, the Netherlands, Denmark and Italy, contributed 73% (0.89 million tonnes) of the total mussel landings, a situation that was much the same 10 years previously (Table 7.1). The common or blue mussel, *Mytilus edulis*, and the Mediterranean mussel, *Mytilus galloprovincialis*, both widely distributed species, form the bulk of total world production.

The common mussel is found in boreal or temperate waters of both the northern and southern hemispheres. In Europe its range extends from the White Sea, Russia and Iceland as far south as the Atlantic coast of southern France. In the northern hemisphere it occurs on the east coast of North America from the Canadian Maritimes southwards to Cape Hatteras, North Carolina, and in the southern hemisphere in Argentina and Chile, the Falkland Islands and the Kerguelen Islands. The Mediterranean mussel is also found in temperate waters, but its range extends into warmer waters. The systematic status of the two species has been subject to much debate, but recent studies (see review by Gosling, 1992) suggest that the Mediterranean mussel is distinct from the common mussel. Its range in Europe extends from the British Isles to the Iberian Peninsula and into the Mediterranean Sea. Elsewhere in the northern hemisphere, it occurs in southern California, Japan, Hong Kong and along the east China coast. In the southern hemisphere, it occurs in Western Australia, Tasmania, New Zealand and South Africa.

The green shell mussel, *Perna* spp., a tropical/sub-tropical genus, is making an increasingly important contribution to some countries' cultivation statistics. In the Far East (Indonesia, Malaysia, the Philippines, Singapore and Thailand) and India, *P. viridis* is cultivated on poles, rafts and longlines. In Venezuela, Ecuador, Brazil, and South Africa, *P. perna* is cultivated on rafts and longlines. However, the species which has shown the most impressive production statistics, is *P. canaliculus*, the New Zealand green shell mussel, whose landings, using modern longline cultivation methods, have increased in value from 1348 t to 50 000 t in the 13 year period to 1994.

Table 7.1 World mussel landings (FAO, 1999)

Species	Common name	Country	Thousands of tonnes		
			1987	1997	10 year average*
Mytilidae		China	312.7	398.2	447.2 (67%)
Mytilus edulis	Common/blue mussel	Spain	247.6	188.8	180.2 (105%)
		Netherlands	98.4	93.2	82.7 (113%)
		Denmark	85.7	90.8	104.8 (87%)
		France	52.3	52.5	51.0 (103%)
		UK	5.2	32.1	12.0 (268%)
		Germany	25.9	22.3	28.3 (79%)
		Ireland	14.9	18.0	15.4 (117%)
		USA	21.5	9.4	18.9 (50%)
M. galloprovincialis	Mediterranean mussel	Italy	85.4	124.4	104.4 (119%)
		Greece	1.3	30.7	11.4 (267%)
		France	4.0	12.1	13.4 (90%)
		Turkey	8.0	8.4	6.8 (125%)
M. viridis	Brown mussel	Singapore and Malaysia	1.6	4.6	2.7 (167%)
M. coruscus	Korean mussel	Korea	29.8	66.8	36.1 (185%)
Perna canaliculus	New Zealand mussel	New Zealand	17.7	65.5	40.0 (164%)
M. chilensis	Chilean mussel	Chile	9.0	13.4	9.4 (141%)
M. smaragdinus	Green mussel	Thailand,	46.8	53.8	57.0 (94%)
		Philippines	13.2	11.7	19.0 (61%)
M. planatulus	Australian mussel	Australia	2.6	1.6	2.2 (73%)
	World total		1077.1	1297.7	1329.3 (106%)

* 1997 values as % of 10 year average (1987–96) are shown in parenthesis

Mussels in general have a number of attributes that contribute to their success as one of the most common of the bivalve species. High fecundity and free-swimming larvae ensure wide distribution of the offspring. Mussels are commonplace on the substructures of offshore oilrigs in the North Sea, 100 km or so from the nearest mussel populations. They settle on rocky shores and intertidally and subtidally in estuaries and bays, often at very high densities. Their rapid growth rate ensures that they are not out-competed by other benthic communities with which they are in competition. Although prone to certain predators (e.g. crabs, starfish and fish: see Chapter 10), mussels are relatively free from the effects of pests and diseases that plague other commercial bivalve species such as oysters.

Perhaps their greatest attribute from the cultivation viewpoint is the mechanism of byssal attachment. The byssus threads, which firmly and flexibly anchor the mussel to its substrate, can be detached and secreted afresh to give new points of attachment. This enables the mussel to reposition itself to gain better access to the water current by obtaining a more favourable position for feeding or extracting oxygen. This is of great benefit for the mussel when it loses its attachment during storms and heavy wave action, or when it needs to re-establish its position during competition from its neighbours either on natural or cultivated beds or when suspended on ropes from rafts, longlines or other structures.

The early life history of the mussel is fairly complex and the description that follows has features that are common to the other mussel genera that are cultivated commercially. Although the sexes are separate, they only become visually distinguishable as they ripen by the creamy coloration of the male and the orange coloration of the female mantle tissue. At spawning, eggs and sperm are released into the sea where fertilisation takes place. Development of the eggs into free-swimming, shelled veliger larvae is quite rapid, usually taking 1–2 days depending on sea temperature. The swimming organ or velum, a ring of cilia that beat vigorously to keep the larva in suspension, is located near the mouth and also creates feeding currents that direct microscopic sized algae towards the mouth. The availability and abundance of the right sorts of algae largely determine the success with which the larva is able to complete this distributive phase of its life.

The larvae remain planktonic for 1–2 months and may drift considerable distances from the parent stock. During this time, the larva increases in size (from 0.1 mm to *c.* 0.25 mm in shell length) and undergoes some morphological changes in preparation for its sedentary way of life. The velum begins to degenerate and is eventually lost. Meanwhile a muscular foot and a pair of eyespots develop, which enable the pediveliger to explore the sea bed for a surface suitable for settlement. Certain algae and hydroids at extreme low water and below are especially attractive for settlement. These primary spat attach themselves by secreting their first byssus threads and remain attached until they reach 1–1.5 mm in size. At this size, they detach and undergo further distribution in water currents until they finally settle on established mussel beds, stones, rocks, shells, mooring chains and other surfaces with small crevices and cracks which offer them shelter and protection. These secondary spat may settle in huge densities, often as high as 100 000 spat per m^2 (Dare, 1980).

Methods of cultivation

Over 40 countries world-wide are listed as significant producers of mussels (FAO, 1999). Most of the production involves some method of cultivation which, when categorised, can be reduced to either two-dimensional bottom culture or three-dimensional suspended culture. Of the methods of cultivation listed, bottom cultivation represents 15% and off-bottom (raft, longline, pole, rack etc.) 85% of the total. The choice of method reflects tradition, site availability, site suitability and availability of seed.

Bottom cultivation

Bottom cultivation is largely a European practice with considerable quantities of mussels being produced annually by the Netherlands and Germany and smaller quantities by Ireland and the UK. France also has substantial bottom mussel production but mainly from natural beds in Normandy [50 000 t in 1993 (Goulletquer & Héral, 1997)]. The occurrence of sheltered, intertidal and subtidal areas with abundant and regular spatfalls is the principal reason for strong mussel production.

The Wadden Sea, a sheltered, shallow body of water off the coast of the Netherlands, Germany and Denmark, is a highly productive nursery ground not only for mussels but also for cockles, shrimp and flatfish. Mussel cultivation in the Wadden Sea is the preoccupation mainly of the Dutch, who have leased substantial sectors of the western end to mussel cultivators. The Oosteschelde, the Netherlands, also contains important mussel cultivation grounds, but the continuity of this fishery is dependent on an annual seed supply from the Wadden Sea.

Similarly productive grounds occur in England, with The Wash on the east coast and Morecambe Bay on the west coast traditionally having provided the bulk of mussel landings. In recent years, especially in The Wash, spatfalls have failed consistently and have led to negligible production of mussels in the late 1990s. Despite this bleak picture of the traditional English fisheries, mussel cultivation has blossomed elsewhere in the UK. In Poole Harbour on the south coast, and in the Menai Strait, north Wales, cultivation has grown steadily in the 1990s as the direct result of re-laying locally collected seed on leased plots, reaching an annual combined production of *c.* 10 000 t in 1997.

Bottom culture is based on transferring mussels from natural, frequently exposed beds to sheltered culture plots where the density is reduced to improve growth and fattening (Fig. 7.1). One-year-old seed mussels, 10–30 mm in size, are ideal for re-laying, but scarcity of stocks may dictate that older, slow-growing mussels from high-level lays may have to be used. For best results, mussels are re-laid as near to low water of spring tides as possible, preferably subtidally. The seed mussels are usually dredged for re-laying in spring and early or late summer, with the cultivation cycle taking 1- 2½

Fig. 7.1 Dredging seed for re-laying in the River Conwy, Wales. (Photo: D. B. Edwards)

years depending on location and market size. Stocking densities of 25–30 t per ha are used for best results. However, yields very much reflect initial stocking density and the care and attention that the seed receive at re-laying. The Dutch experience shows that, with careful re-laying, yields of 3 t of market size mussels per t of seed may be obtained, but this may decline to 0.5 t per t re-laid if poorly managed, with high losses due to drift, storms, predation etc. The average yield is likely to be nearer to 1:1. However, experimental studies in Wales, using crab-proof, intertidal, fenced plots, showed that productivity could be considerably higher, in the region of 8:1, with the exclusion of the main predator, the shore-crab, *Carcinus maenas*, and careful husbandry (Dare *et al.*, 1980) (see Chapter 10).

The methods employed for harvesting mussels may range from simple hand-gathering and hand-raking on low-yield fisheries such as the Conwy fishery in north Wales (Fig. 7.2) to power dredging on high-yield fisheries. The Dutch industry uses large, powerful dredgers deploying four dredges, each with a capacity of 0.5 t and capable of harvesting more than 100 t of mussels in a day's fishing. In Poole Harbour, England, new technology developed by the Sea Fish Industry Authority to reduce handling and costs, has been installed in several of the mussel-harvesting vessels. These deploy a single, self-emptying dredge, which harvests several hundred kilograms of mussels every few minutes from the culture plots. The catch is released into

Fig. 7.2 Gathering mussels by hand-raking in the River Conwy, Wales. (Photo: D. B. Edwards)

a hopper, then fed into a de-clumping, washing and grading machine before being bagged. The containers, either 20 kg plastic mesh sacks or 1 t canvas bags, are a convenient way of transferring the catch ashore for further processing.

Off-bottom culture

World-wide, the vast majority of cultivated mussels are grown above the sea bed. Although the reasons chosen for doing this may reflect local availability of sites and materials, the common advantages are that the crop is less accessible to benthic predators and that it has better access to food-bearing water currents.

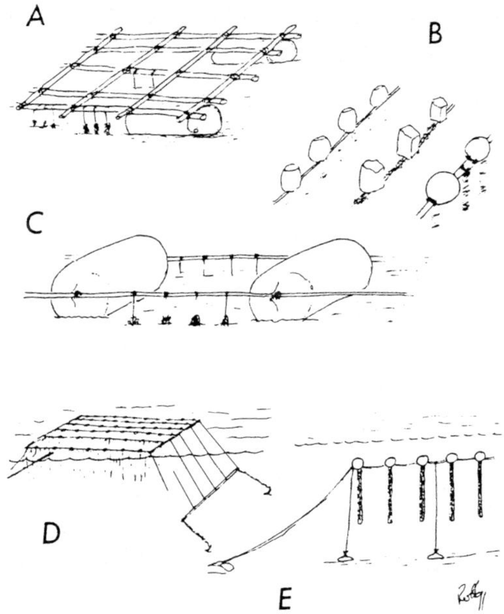

Fig. 7.3 Various methods of floating suspended cultivation. (A) bamboo and oil-drum raft, (B) single longline, (C) double longline, (D) multi-longline, (E) subsurface longline. (Adapted from Hickman, 1992)

The three principle methods of off-bottom culture, namely pole, raft and longline, have many variants (Figs 7.3, 7.4) that have arisen from local needs to use particular materials to keep the costs of cultivation to a minimum.

Pole culture

This method of cultivation, using poles or stakes driven into the sea bed (Fig. 7.4*)*, is essentially used in shallow water with access mainly at low tide and in some circumstances by diving. It is probably the oldest method of mussel cultivation in Europe, its origins beginning in France in the thirteenth century where it is known as 'bouchot' culture. In 1993, French production from 1613 km of bouchot poles amounted to 58 000 t, a yield greater than that from either of the ground or longline methods of culture used in that country.

The bouchot culture method, which has changed little since its origin, consists of inserting intertidally rows of vertical poles in the sea bed. The poles, made of pine, oak, Brazilian hardwood or occasionally aluminium, are 4–7 m long and 25–30 cm in diameter and are inserted in the soil so that *c.* 2–3 m is above the sea bed. They are placed 25–50 cm apart, and are set

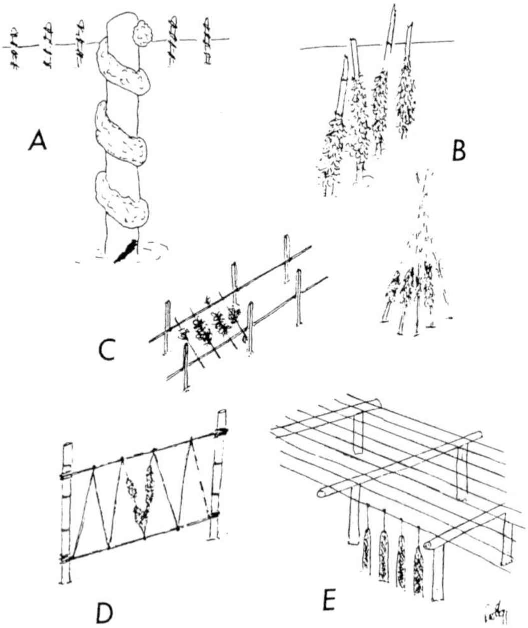

Fig. 7.4 Various methods of fixed suspended cultivation. (A) bouchot, (B) bamboo pole, (C) rack and rod, (D) rope-web, (E) hanging park. (Adapted from Hickman, 1992)

closer together for spat collection or further apart for growing to market size. Depending on location, the rows may comprise up to 250 poles and are positioned near to low water mark of spring tides, in parallel lines at right angles to the axis of the beach. The poles, especially those used for spat collecting, are set out early in the year. The south-western coast of France produces regular settlements of spat in the spring (May–June) and these are used for seeding other areas including the less productive areas further north (e.g. north Brittany and Normandy). Spat are collected on poles or horizontal lines of coir rope on metal frames placed in deeper water. In July, the seed are transferred to tubular nets, 3–5 m in length, which are wound round and nailed at each end to the growing poles. The seed migrate out from the tube and spread over the surface of the pole. In some cases, the mussels may become overcrowded and need thinning during the summer. This is achieved by placing the surplus mussels into fresh tubes on other poles. At the end of the cultivation cycle, which may take up to 2 years, a pole may produce 25–60 kg of mussels, 40–50 mm in size.

Raft culture

With this method of culture, the mussels are grown in a three-dimensional matrix that concentrates them into a relatively small area. Since suspended ropes are used as the medium for attachment of mussels, the method essentially requires a depth of water several metres greater than the length of the ropes to prevent them grounding at low tide and allowing access by predators. Other practical aspects, such as shelter, current speed and water quality, are prime considerations for the welfare of the mussels. In recent years poorly sited raft and suspended rope farms have caused environmental damage to natural populations of bottom-living animals and plants owing to the deposit of smothering, anoxic layers of mussel faeces and pseudofaeces, which collect on the bottom.

Spain, the largest producer of mussels in Europe with an average annual crop of *c.* 150 000 t (248 000 t in 1987; 189 000 t in 1997; FAO, 1999) employs rafts as the major cultivation method. It has a very young history, beginning in the 1940s, and is centred in the deep, sheltered, productive waters of the drowned river valleys (rias) of the Galician coast of north-west Spain. Early rafts used old ships' hulls as a means of flotation. The hull supported timber spars across its beam from which ropes were suspended. Such a raft would also carry a platform and shelter for the workmen and bear one or more masts with stays connecting the top to the extremities of the timber beams to support the heavy load of mussels.

With time, specialised rafts were developed, with purpose-built flotation comprising plastic or wooden floats encased within a thin outer concrete or glass fibre layer to protect them from attack by marine wood-boring organisms. The framework of parallel rows of timber beams above the floats was made from readily available eucalyptus timber. A typical raft, 20 m square, carrying *c.* 500 ropes spaced 50 cm apart on wooden beams, also *c.* 50 cm apart, might produce up to 60 t of mussels per year (Mason, 1972). By 1995 3386 rafts were employed in this industry producing 230 000 t of mussels. Recent Spanish statistics show that average production is *c.* 130 kg per metre^2 of raft area (10 kg/m of rope) and 20–100 t for an entire raft, depending on it's size, with a mean value of 47 t (in Caceres-Martinez & Figueras, 1997).

Raft cultivation has become popular in many countries, notably Australia, China, Chile, Canada, USA, India, Ireland, Malaysia, New Zealand, Scotland, Singapore and Venezuela. The level of technology varies considerably from simple flotation units bearing ropes of natural fibres to structures employing the latest ideas in engineering and materials that are designed for longevity in the harsh marine environment. Flotation is important, with units ranging from empty air-filled plastic and metal drums to purpose-made expanded-foam-filled containers of a size appropriate to the carrying capacity of the raft.

Although mussel cultivation in Scotland is still in its infancy (1000 t in 1996), the industry is applying modern technology to the use of both raft and longline culture. Raft design ranges from imported Spanish equipment to units purpose-built in Scotland (Fig. 7.5). The Spanish raft, measuring 27

Fig. 7.5 Raft used for mussel cultivation, Scotland. (Photo: Xplora Products Ltd.)

× 20 m, is made of eucalyptus timber from which 850 synthetic ropes, each 8–10 m long, are suspended. Flotation is provided by four large steel drums. The raft is attached to the sea bed by a single point mooring which allows it to swing with the tide. The Scottish-made Muckairn raft is used extensively on the west coast of Scotland. The plastic foam-filled floats (120 × 80 × 65 cm) support wooden beams 11 m long (10 × 8 cm), spaced 50 cm apart, on galvanised tubular joists (10 × 10 cm). Two hundred synthetic ropes (8 m long × 16 mm diameter) are attached to the beams at 25–40 cm intervals. To reduce the likelihood of crop loss from slippage, 25 cm-long wooden pegs are inserted through the rope every 25–40 cm. The raft has a carrying capacity of more than 10 t (*c.* 6 kg per metre of rope). Eight rafts, attached in series, are anchored by eight 3 t concrete mooring blocks. The raft costs $US10 000–11 000 and produces a crop valued at $US16 300 (Karayucel, 1997).

Longline culture

Although this method is more suitable than rafts for cultivation at exposed sites, the two methods are frequently found together in sheltered waters. Modern systems are adaptations of methods developed in Japan for cultivation of oysters and scallops. Typically, it consists of a series of horizontal

Fig. 7.6 Double longline, Marlborough Sounds, New Zealand.

polypropylene ropes buoyed at the surface or, in some circumstances, 1.5–3 m below the surface to avoid surface layers of fresh water. Dropper ropes attached to the headlines at *c*. 50 cm intervals carry the mussel crop. In some circumstances where current velocity is sufficiently high, the ropes may require weighting at the ends to prevent them swinging into each other. A longline 200 m in length uses 18–30 mm diameter rope which is supported by 25–301 plastic floats, 0.5–1.5 m apart depending on the carrying capacity of the system. Extra floats are required as the mussels grow towards market size to prevent the crop from touching the bottom. The vertical polypropylene ropes carrying the mussels are usually 4–6 m long and 14–18 mm in diameter. As with raft culture, wood or plastic pegs are inserted at intervals through the rope strands to stop the mussels from slipping off.

Double longline systems, popular in New Zealand and newly introduced into Scotland, have the advantage of using single 1 t mooring blocks at each end. The two surface lines of ropes are buoyed at intervals by special plastic floats interconnecting them (Fig. 7.6). The floats fitted with a notched handle on each side into which the rope is tied, allows them to be quickly and easily added or removed from the system. The physical carrying capacity of a longline depends on the volumetric capacity of the floats. Assuming the floats are of similar size and with 70% of their capacity providing lift, 50 floats, each 250 l in volume could support 35 000 kg of mussels. This is calculated by:

$$50 \times 250 \times 0.7/0.25^{\#}$$

with the denominator (#) applied to obtain the weight of mussels in sea water (see Jenkins, 1985). Whether the number of floats required to support the crop weight at harvesting are added at the beginning of the cultivation cycle or progressively as the mussels grow towards marketable size, is a question of economics, management strategy and other physical and biological factors for the farm manager to consider. An example of the carrying capacity of dropper lines of various lengths, and the buoyancy required to support green shell mussels is given in Table 7.2.

A typical New Zealand double longline system might be 100 m long and carry single droppers 8–10 m long or more recently a series of continuous loops, each 25 m long, which are attached alternately to the pair of parallel lines. A farm might be 3 ha in area with four rows of buoys each carrying 100 m longlines running parallel to the shore. The rows are set along the flow of current, with the first row 50 m and the fourth row 200 m away from the shore. The buoys are positioned sufficiently far apart to allow a boat easy access for servicing and harvesting the crop.

In some countries, subsurface longlines (Fig. 7.3) are used to enable cultivation to take place in offshore areas exposed to excessive wind and wave action or, as in the case of Canada, to avoid them becoming embedded in sea ice in the winter. This increases the options for cultivation for some countries with a long, exposed coastline. France, for example, produced 30 000 t of mussels in 1993 by this method, with production rates reaching 18–20 t per ha per year, on its Atlantic and Mediterranean coast.

Table 7.2 Weight of marketable green shell mussels (average length of 100 mm and live weight of 80 g) on dropper lines of various lengths, and buoyancy required to support stock at 250 mussels per metre (derived from data in Hickman, 1992)

	Crop weight (kg)		
Length of dropper (m)	In air	In water	Buoyancy required (l)*
5	100	25	35.7
8	160	40	57.1
10	200	50	71.4

* Assuming 70% of the volume provides the lift required to support crop weight in water

Spat collecting

Spat collecting plays an important part in the continuity of suspended mussel cultivation. Although some countries, e.g. China and Thailand, use hatchery techniques for some of their spat collecting, it is an expensive method for a relatively low-value species. However, with the increase in value of some species of mussels (e.g. green shell mussel in New Zealand and *M. galloprovincialis* in the USA), profitability of hatchery production is being re-examined. In the main, spat collection depends on gathering wild spat, and in circumstances where spat collectors are deployed requires a fairly accurate knowledge of the timing of spatfall. In temperate waters, mussel spawning/settlement usually occurs at well-defined times of the year, whereas in tropical/subtropical areas it may occur regularly, with perhaps less well-defined peaks, over much of the year. Vagaries of nature, however, frequently interfere with accurate forecasting, with heavier or lighter and earlier or later levels of spatfall than expected occurring. A manager's experience of prevailing conditions would enable him to judge the timing and quantities of collector required to cope with normal variation of spat production and to spread the risk by collecting spat at more than one site of collection.

Materials used for spat collection vary widely and depend on their suitability as collectors, local availability, durability and cost. In Thailand and the Philippines bamboo and wooden stakes, similar in principle to the French bouchot poles, are used. Ropes are one of the most commonly used collecting materials because of their ease of use and relative cheapness. Ropes of natural fibres, such as coir (coconut fibre) and sisal, and synthetic materials such as polypropylene and polyethylene, are popular for spat collection. Natural fibres are especially good since the hairy nature of their surfaces is very attractive to mussels for settlement. Natural fibres, although quickly degraded in the marine environment, are the main collecting and grow-out materials in some countries because of their cheapness and ready availability. Synthetic ropes, however, are the most commonly used materials for mussel culture in developed countries. Their suitability for spat collecting can be improved considerably by inserting 15 cm long sections of unravelled rope at 10 cm intervals [split-film polyethylene (SFP) in Fig. 7.7 and with further reference in Fig. 7.8 and Table 7.3] to create the numerous crevices which enhance spat settlement.

Collector ropes are suspended from longlines or rafts in much the same manner as the culture ropes. In Atlantic waters of Europe, although subject to some geographic variation, spatfall usually occurs in June–December following major spawning in February–May. In New Zealand, spatfall occurs in February–April. The timing of deployment of collectors may be helped by hanging small sections of rope at different depths of the water at the collecting site as spatfall indicators. Frequent examination of these test ropes then provides a clue for the deployment of the main collectors. Some examples of settlement densities of mussels in different parts of the world (Table 7.3) and using different collecting materials (Fig. 7.8) show the range that may be expected. Season, type of collector and its location clearly have an

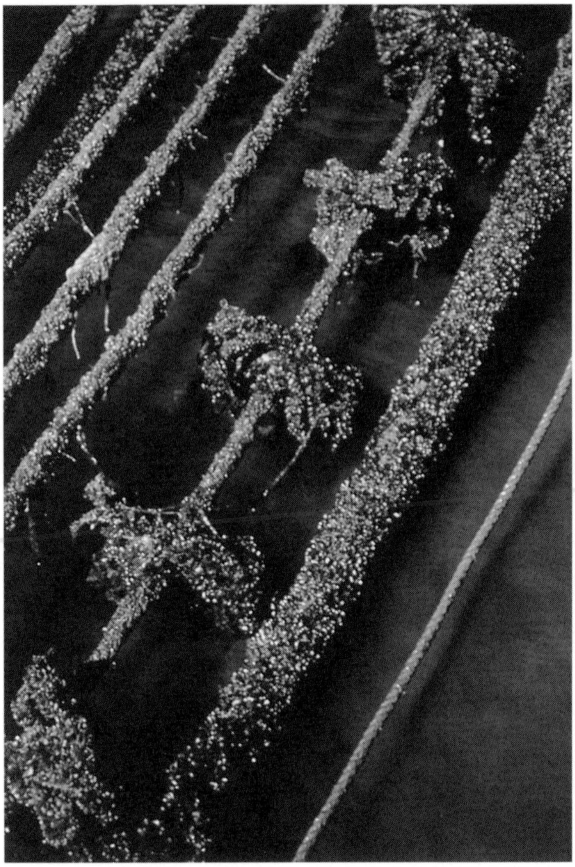

Fig. 7.7 Mussel spat on coir and synthetic (split-film polyethylene) ropes, Morecambe Bay, England. (Photo: D. B. Edwards)

effect on spatfall intensity (Fig. 7.8, from Dare *et al.*, 1983), with maximum values exceeding one million of spat per metre of split film polyethylene rope in the peak period of settlement in this estuary in England. Spatfall densities, however, reduce with time as competition and perhaps predation, take their toll.

Spat are usually removed from the ropes when they reach *c.* 10 mm in size. Placing mussel ropes in sacks in the sea for 24 hours can, in some areas, aid stripping by forcing the spat to loosen byssal attachment. Clumped groups of mussels may be separated mechanically or by hand into individuals. Depending on country or area, the seed are placed in cotton mesh (e.g. New Zealand, France and India) or synthetic mesh (pergolari mesh, Italy) stockings fitted over a culture rope. In Spain, clumped seed are bandaged to

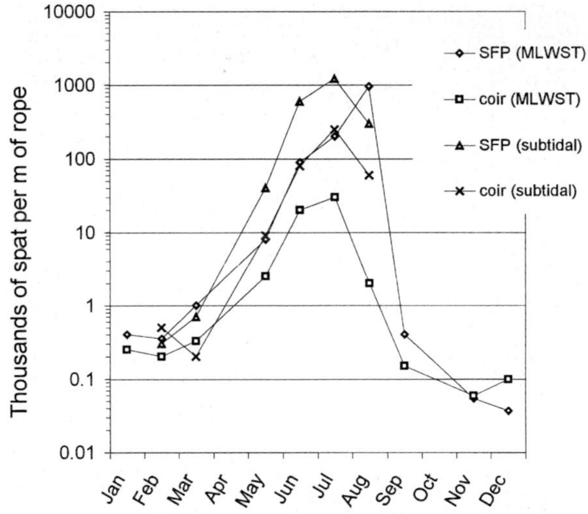

Fig. 7.8 Seasonal spatfall of *Mytilus edulis* on experimental ropes in the River Teign, England, in 1978. (Redrawn from Davies *et al.*, 1980)

Table 7.3 Some spat densities recorded for *Mytilus* and *Perna* spp

Location	Species	Number	Collector	Author
Singapore	*P. viridis*	5000–10 000	Per m^2 of collector surface	Vakily (1989)
England (Morecambe Bay)	*M. edulis*	500–20 000	Per m of coir rope on and off ground	Dare, Davies & Edwards (1983)
"	"	1800–140 000	Per m of split film polypropylene off ground	"
"	"	2000–55 000	Hairlock on raft	"
China	"	30–929	Per m of grass rope	Zhang (1984)
New Zealand	*P. canaliculus*	up to 20 000	Per m of rope on longlines	Jenkins (1985)

the rope with a fragile cotton mesh tape. Within a few weeks, the mesh degrades leaving the spat firmly attached to the rope by the byssus, or the spat are able to force their way through the synthetic mesh to gain access to the water supply.

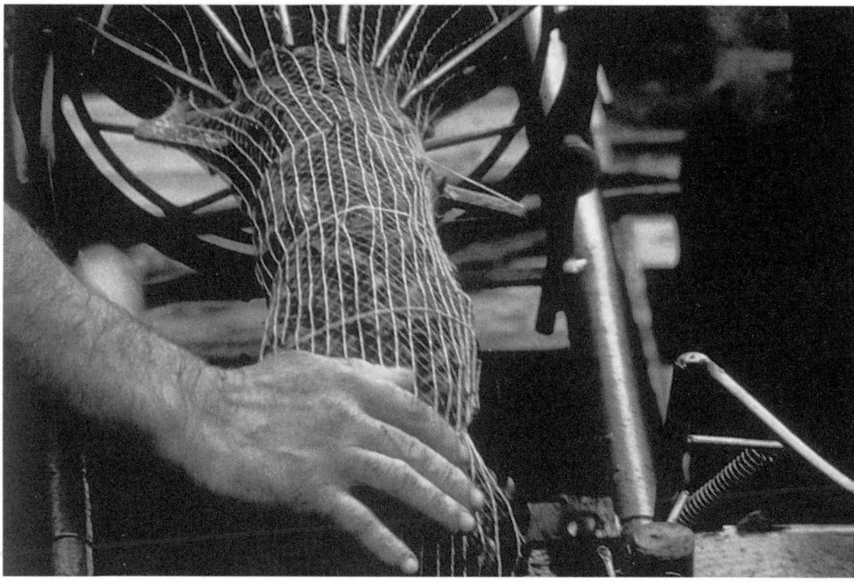

Fig. 7.9 Sleeve for feeding mussel spat into mesh stockings. (Photo: Sea Fish Industry Authority).

Stocking densities depend on the size of the species composing the final crop. In Spain, stocking densities range from 6000 10 mm seed to 300 40 mm seed per metre of rope for *M. galloprovincialis*; in New Zealand, 200–300 10 mm seed per metre for *P. canaliculus*. In Ireland, rates ranging from 0.5 to 1.5 kg per metre of rope are quoted for *M. edulis* up to 20 mm in size. At any given size, seeding density can be critical. For example, a 25% difference above or below a seeding rate of 1.5 kg/m for 20 mm seed can lead to overcrowding or understocking a rope (G. Davies personal communication, 1998). The stocking procedure is simplified by feeding the mussels through a plastic sleeve to keep the mouth of the stocking open (Fig. 7.9).

Some culture areas may be unsuitable or unreliable for spat collecting. For example, spat destined for rope culture in the Spanish rias are collected from rocks on the Atlantic coast. In New Zealand, although some farms set out special ropes for collecting spat at their site, the industry is reliant on mussel spat washed up on the beach attached to seaweed at Ninety Mile Beach, North Island. These spat, known as 'Kataia' spat, a few millimetres in size when collected, may be grown on ropes by specialist nursery growers for later resale or sold directly to farms for rope cultivation. Although the timing of arrival of the Kataia spat is unpredictable and usually different to that of local spat, it usefully allows restocking to take place over a longer period of the cultivation cycle.

Table 7.4 Yields of off-bottom cultivated mussels (from Hickman, 1992)

Species	Country	Yield (kg/m)	Method
Mytilus edulis	Ireland	5	raft
"	France	6–11	bouchot
"	Italy	10–14	longline
"	Sweden	7	longline
M. galloprovincialis	Spain	14	raft
Perna perna	Brazil	6–7	raft
P. canaliculus	New Zealand	20	longline/raft
P. indica	India	16	raft
P. viridis	Singapore	10–12	raft

Growth rates and yields

The range of yields given for mussels cultivated by off-bottom methods (Table 7.4) shows a large variation (5–20 kg per metre of rope) caused by differences between species, location and other biological, physical and husbandry factors.

These factors also affect growth rate, and for a given species there may be size preferences at planting and marketing which influence the duration of the cultivation cycle (Table 7.5).

Prospects

Although world mussel production increased by only 6% in the period 1987–97 (FAO, 1999; Table 7.1), the statistics conceal true growth performance because of hidden annual fluctuations in production levels. Average production over a 10 year period gives a better indication of change in performance. Thus it would seem that total production of mussels in 1997 was near to the 10 year average (Table 7.1) with Korea, Chile, UK, Greece and New Zealand being the only countries showing a significant increase in production.

Although these statistics suggest that world demand for mussels may be near to saturation point, industry growth will be influenced by various factors. These are likely to include:

- exploiting new (e.g. more exposed offshore locations) or more effective use of existing cultivation niches,
- careful use of the seed resource,
- effective management of the environment to ensure that farming practice is ecologically benign,

Table 7.5 Growth of cultivated mussels (from Dare, 1980; Hickman, 1992; Stirling & Okumus, 1995)

Species	Country	Growth	Method
Mytilus edulis	England & Wales	55–57 mm/1 year	sea bed (sublittoral)
"	"	65–68 mm/2 years	"
"	"	45–48 mm/1 year	sea bed (MLWST*)
"	"	55–58 mm/2 years	"
"	"	40–42 mm/1 year	sea bed (MLWNT*)
"	"	48–50 mm/2 years	"
"	Scotland	49 mm/1 year	raft
"	Holland	72 mm/3 years	sea bed
"	France	46–54 mm/12 months	bouchot
"	Sweden	60 mm/18 months	longline
"	Norway	46 mm/3 years	longline
"	Canada	55 mm/18–24 months	longline
"	Canada (British Columbia)	50 mm/12–14 months	raft
"	USA	43–50 mm/ 12–15 months	raft
"	China	70–80 mm/ 12–18 months	raft
M. galloprovincialis	Spain	80–90 mm/ 12–18 months	raft
M. chilensis	Chile	54 mm/14 months	raft
Perna viridis	India	93 mm/12 months	raft
"	Tahiti	90 mm/10 months	rack
"	Philippines	50–100 mm/ 6–10 months	poles
P. canaliculus	New Zealand	70–115 mm/12 months	raft/longline
"	New Zealand	113–165 mm/2 years	raft/longline
Choromytilus chorus	Chile	60 mm/2 years	raft
"		120 mm/6 years	raft
Aulacomya ater	Chile	70 mm/2 years	raft

Data for England and Wales (Dare, 1980) and Scotland (Stirling & Okumus, 1995), are for mussels seeded at 25 mm and 27 mm respectively. * intertidal culture at mean low water of spring tides (MLWST) or of neap tides (MLWNT)

- competition from other water and land users,
- improved marketing strategy with emphasis on value added products and international trade.

References and bibliography

Caceres-Martinez, J. & Figueras, A. (1997) The mussel, oyster and pectinid fisheries of Spain. In: *The History, Present Condition, and the Future of the Molluscan Fisheries of North and Central*

America and Europe, Vol. 3, Europe (eds C. L. MacKenzie, Jr, V. G. Burrell, Jr, A. Rosenfield & W. L. Hobert), pp.165–90. US Department of Commerce, NOAA Technical Report **129**.

Dare, P. J. (1980) *Mussel Cultivation in England and Wales.* MAFF, Directorate of Fisheries Research, Lowestoft, Laboratory Leaflet **50**.

Davies, G., Dare, P. J. & Edwards, D. B. (1980) *Fenced enclosures for the protection of seed mussels* (Mytilus edulis L.) *from predation by shore-crabs* (Carcinus maenas (L). MAFF, Directorate of Fisheries Research, Lowestoft, Fisheries Research Technical Report, **56**.

FAO (1999) *Yearbook of Fishery Statistics for 1997*. Vol. **84**. Food and Agriculture Organisation of the United Nations, Rome, Italy.

Gosling, E. M. (1992) Systematics and geographic distribution of *Mytilus*. In: *The Mussel* Mytilus: *Ecology, Physiology, Genetics and Culture* (ed. E. Gosling), *Developments in Aquaculture and Fisheries Science*, **25**, pp. 1–20. Elsevier, Amsterdam.

Goulletquer, P. & Héral, M. (1997) Marine molluscan production trends in France: from fisheries to aquaculture. In: *The History, Present Condition, and the Future of the Molluscan Fisheries of North and Central America and Europe*, Vol. 3, Europe (ed. C. L. MacKenzie, Jr, V. G. Burrell, Jr, A. Rosenfield & W. L. Hobert), pp 137–64. US Department of Commerce, NOAA Technical Report **129**.

Hickman, R. W. (1992) Mussel Cultivation. In: *The Mussel* Mytilus: *Ecology, Physiology, Genetics and Culture* (ed. E. Gosling), *Developments in Aquaculture and Fisheries Science*, **25**, pp. 465–510. Elsevier, Amsterdam.

Jenkins, R. J. (1985) *Mussel cultivation in the Marlborough Sounds (New Zealand)*. 2nd edition revised by J. Meredyth-Young & G. D. Waugh. New Zealand Fishing Industry Board.

Karayucel, S. (1997) Mussel culture in Scotland. *World Aquaculture*, March 1997, pp. 4–10.

Mason, J. (1972) The cultivation of the European mussel, *Mytilus edulis* Linnaeus. (ed. H. Barnes), *Oceanography and Marine Biology an Annual Review*, **10**, 437–60.

Stirling, H. P. & Okumus, I. (1995) Growth and production of mussels (*Mytilus edulis* L.) suspended at salmon cages and shellfish farms in two Scottish sea lochs. *Aquaculture*, **134**, 193–210.

8 Scallop cultivation

Introduction

Although there are about 400 pectinid species worldwide, only *c.* 33 species are exploited commercially. Five of the 28 European species are commercially important. World scallop production reached 1.7 million tonnes in 1997, having doubled during the previous decade (FAO, 1999; Table 8.1), mainly owing to increased landings by Japan and China. Production in these two countries far outstrips that of any other. Although *Patinopecten yessoensis*, the Japanese scallop, is the major species of world landings, FAO data hide the fact that other scallop species, such as *Argopecten irradians* (a non-native species introduced in 1982) and *Chlamys farreri*, contribute significantly to the landings of China.

About three-quarters of world production is achieved by aquaculture. The only significant fisheries species is *Placopecten magellanicus* (7%) of the north-west Atlantic seaboard of Canada and USA.

History of cultivation

Scallop cultivation is a relatively new activity, beginning in Japan in the 1930s. Even so, there was a lag in wide-scale application of cultivation techniques in Japan until the 1960s, when production from wild fisheries began to decline severely owing to overfishing. The meteoric success of cultivation since that time is shown in their landings statistics. In 1965 production from the wild fishery was at a low of *c.* 10 000 t per year. This provided the stimulus for a concerted effort to expand production by intervention in the natural process. By 1975 production had risen to 110 000 t, and by 1997 to ~500 000 t (FAO, 1999), mainly from cultivation.

The focus of attention was the Japanese scallop (*Patinopecten yessoensis*), a cold-water species occurring in the north of the country, in Hokkaido and along the northern coasts of Honshu. In the early days it was found that spat could be collected on scallop shell and foliated cedar twigs suspended in the water column. However, in recognition of the behavioural response that causes young spat to detach from suspended collectors in search of the

Table 8.1 Main areas of scallop production (FAO, 1999)

Species	Common name	Country	1987	1997	10 year average*
			Thousands of tonnes		
Patinopecten (Mizuhopecten) yessoensis	Japanese scallop	China	43.6	1001.5	443.9 (226%)
		Japan	297.8	515.2	417.5 (123%)
Pecten maximus	Scallop	Europe (mainly France, UK, Isle of Man)	15.5	33.6	18.8 (179%)
Pecten novaezelandiae	New Zealand scallop	New Zealand	0.9	2.6	1.0 (253%)
Placopecten magellanicus	American sea scallop	Canada	73.8	53.7	78.1 (69%)
		USA	119.7	48.1	96.5 (50%)
Patinopecten caurinus	Weathervane scallop	USA	2.7	<0.01	3.4 (<0.01%)
Argopecten gibbus	Calico scallop	USA	85.4	16.8	61.7# (27%)
Argopecten irradians	Bay scallop	USA	2.9	0.5	1.7 (50%)
Argopecten purpuratus	Peruvian calico scallop	Chile, Peru	5.6	15.8	7.4 (213%)
Chlamys islandica	Iceland scallop	Mainly Iceland	13.4	10.6	10.6 (100%)
Chlamys (Aequipecten) opercularis	Queen scallop	Europe (mainly Faeroe Islands, France, UK, Isle of Man)	14.3	11.2	13.1 (86%)
	World total all species		744.9	1746.3	1182.4 (148%)

* 1997 values as % of 10 year average (1987–1996) are shown in parenthesis. # data for six years only

bottom sediment, the collectors were enclosed within small-meshed net bags to prevent loss of spat. After a period of time, when the spat had grown to 5–10 mm, they were transferred to suspended nets for further growing to a size suitable for relaying on the ground (30 mm) or through to marketable size of *c.* 100 mm. Depending on area and method of cultivation, the cycle takes 1½–3½ years to complete.

Without doubt Japan has been the cradle of invention of modern scallop cultivation. The methods developed there have been applied in many countries world-wide. Japan and China continue to dominate world landings, contributing, mostly through aquaculture, 86% of the total in 1995. Chile has also made significant progress in scallop cultivation using mainly hatchery technology to produce the seed for on-growing. In New Zealand

Table 8.2 Culture times (from spat to harvest) of pectinids (adapted from Dare, 1996)

Country	Species	Minimum landing size (mm)#	Culture period (years)	Annual temperature range (°C)	Area
China	*Argopecten irradians*	55–70	<1	−1–30	Gurei Bay (suspended)
	Chlamys farreri	50–70	2–3	4–22	
	Patinopecten yessoensis	70	?	?	
Japan	*Patinopecten yessoensis*	100 (L)	2½–3½	5–9	Hokkaido (bottom)
	" "	100 (L)	1½–1¾	5–25	Mutsu Bay (suspended)
	Mimachlamys senatoria nobilis	70 (L)	1½		
Russia (east)	*Patinopecten yessoensis*	100 (L)	3–5	−2–26	Vladivostok (bottom)
Australia	*Pecten fumatus*	80 (L)	1¼–2	?	Tasmania (suspended)
New Zealand	*Pecten novaezelandiae*	100 (L)			Tasman Bay (bottom)
Chile	*Argopecten purpuratus*	80–90 (L)	1–1½	17–19	N. Chile (suspended)
Mexico	*Argopecten circularis*	~80 (L)	<2	19–30	Baja California (suspended)
USA (Pacific)	*Crassodoma gigantea*	100 (H)	2		California
Canada (Pacific)	*Crassodoma gigantea*	100 (H)	3–3½		Nanaimo (suspended)
	Patinopecten yessoensis	100 (H)	2		" "
Canada (Atlantic)	*Placopecten magellanicus*	90 (H)	2¾–3		New Brunswick (suspended)
	Placopecten magellanicus	90 (H)	4–5		" (bottom)
USA (Atlantic)	*Argopecten irradians*	50	<1	0–29	SE States (suspended)
	Argopecten gibbus	40	<1		" "
Argentina	*Chlamys tehuelcha*	60 (H)	2¼		(suspended)
France	*Pecten maximus*	100 (L)	3½–4	8–16	Brittany (bottom)
UK	*Pecten maximus*	100 (L)	4–5	6–16	W Scotland (bottom)
	Chlamys opercularis	none	~2	"	" (suspended)
Italy	*Pecten jacobaeus*	105 (L)	1½	9–28	N Adriatic (suspended)

# denotes shell length (L); shell height (H)

abundant natural spatfalls in Nelson Bay, collected on spat collectors and relaid on designated areas of the sea bed to supplement the ailing dredge fishery, contributed an estimated 40–50% of the landings (700 t meat weight) in 1992. Around 15 other countries have spent up to 20 years developing cultivation techniques without much success. Of these, France, Scotland and Canada, each producing <200 t per year through aquaculture, have turned to hatchery seed production to supplement or replace that supplied irregularly from the sea.

The cultivation period, from spat to harvest, varies widely according to species, temperature, food supply, density, growing systems and minimum legal size (Table 8.2). There is a marked difference between the smaller and faster-growing, warm-water pectinids, some being harvestable after only 1 year (*Argopecten* spp.) and the larger, temperate species (*Pecten*, *Placopecten*, *Crassodoma*), which may require 3–5 years. For the latter group, cultivation may be reduced by up to several years by the use of suspended cultivation, but at a heavier cost. The economic trade-off between the high cost and high production of suspended cultivation and low cost and low production of sea bed cultivation requires careful assessment. Only a few countries, such as China and Chile, with their low labour and equipment costs, are able to sustain a profitable cultivation industry based totally on suspended cultivation. The others are seeking alternative strategies in attempts to harness these lucrative species by cultivation.

Cultivation in various countries

Japan

Although the major component of Japanese scallop landings is *Patinopecten yessoensis* from the north, there is interest in cultivating two temperate water species, the noble scallop [*Chlamys (Mimachlamys) senatoria nobilis*] and the bay scallop (*Pecten albicans*) in southern parts of the country.

Patinopecten yessoensis has long been of commercial interest in Japan. It is a cold-water species occurring naturally in the north-west Pacific Ocean (above *c.* latitude 35° N) and is cultured mainly around the coast of Hokkaido and the north and north-west coasts of Honshu and along the coast of the North Korean peninsula. It has also been introduced into China and British Columbia, Canada. In Japan, the Japanese scallop is largely cultivated by hanging culture, by reseeding areas of the sea bed, and by harvesting wild stocks. Although considerable experience has been gained with hatchery culture, the abundance of wild seed generally precludes the use of this method for seed production.

Spawning occurs in spring (March–April), and with larval development taking 5–6 weeks, spatfall occurs in April–June depending on area. At this time, the pediveliger larva seeks out and attaches to a suitable substrate by means of its byssus thread and completes metamorphosis into the juvenile or spat stage. In nature, the attachment might be to a hydroid stem or filamentous alga. This substrate, however, is not its permanent home, for after a period of time the spat detaches and falls to the sea bed.

Unlike most other bivalves, the scallop has some powers of movement over the sea bed. It achieves this by clapping its valves together in rapid movements. This forces water into and out of the mantle cavity, creating jets sufficiently powerful to move the scallop, perhaps assisted by water currents, a few metres at a time. This movement might be used as an escape response to predators such as starfish or crabs, to reduce competition from over-dense neighbours or to avoid some other unfavorable conditions. Reseeded commercial plots in some areas are spaced 1½ km apart to reduce the risk of movement of scallops onto neighbouring farms.

Seed collection

This is accomplished in the spring by the deployment of onion bags filled with monofilament plastic netting (Fig. 8.1), a method that is now used extensively world-wide. The collectors are suspended from longlines just prior to the expected time of spatfall. This timing is aided by monitoring water temperature, scallop gonad index and number and size of the larvae. Imminent spatfall is expected when 50% of the larvae in a sample are larger than 200 μm. The intensity of settlement is variable, ranging from a few hundred to *c.* 10 000 spat per bag. A few thousand spat per bag is considered ideal, since too many causes reduced growth and survival and too few are an inefficient use of the resource. The quantity of filling material must allow good water penetration to the inside of the bag to ensure that larvae

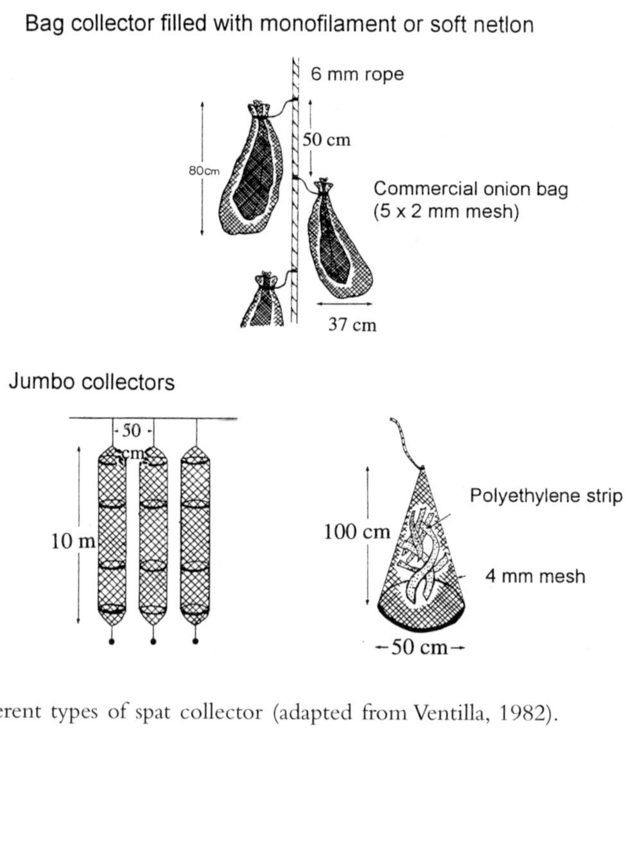

Fig. 8.1 Different types of spat collector (adapted from Ventilla, 1982).

settle successfully and that food subsequently reaches the metamorphosed spat. On average, about 100 g of monofilament filling is used for a bag collector 80 × 37 cm in size. In areas with expected high spatfalls, jumbo collectors (Fig. 8.1) with a sparse filling are used to collect lower numbers of spat, thereby improving their growth and survival.

The collectors are usually suspended at mid-depth from longlines up to 200 m long (Fig. 8.2). Suspending the collectors on sub-surface buoys reduces the effect of wave motion at the surface, improving spat settlement and survival. Water depth and other environmental conditions prevailing at the site determine the configuration of the longline. There may be 10–20 collectors attached to a string and several thousand collectors per longline.

Suspended culture in nets

The spat are removed from the collectors in the summer (July–September). This is done as quickly as possible, and outside the heat of the day, since

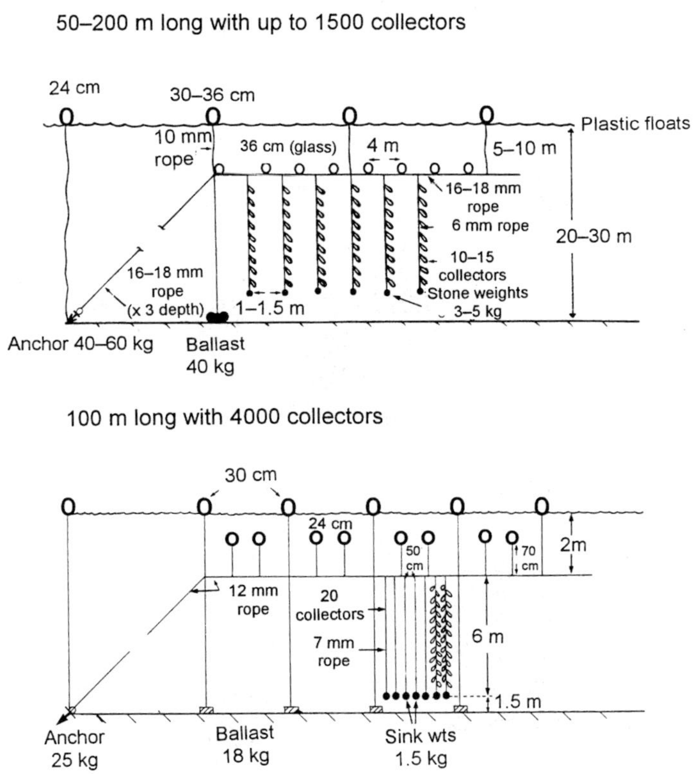

Fig. 8.2 Different methods of deployment of spat collectors (adapted from Ventilla, 1982).

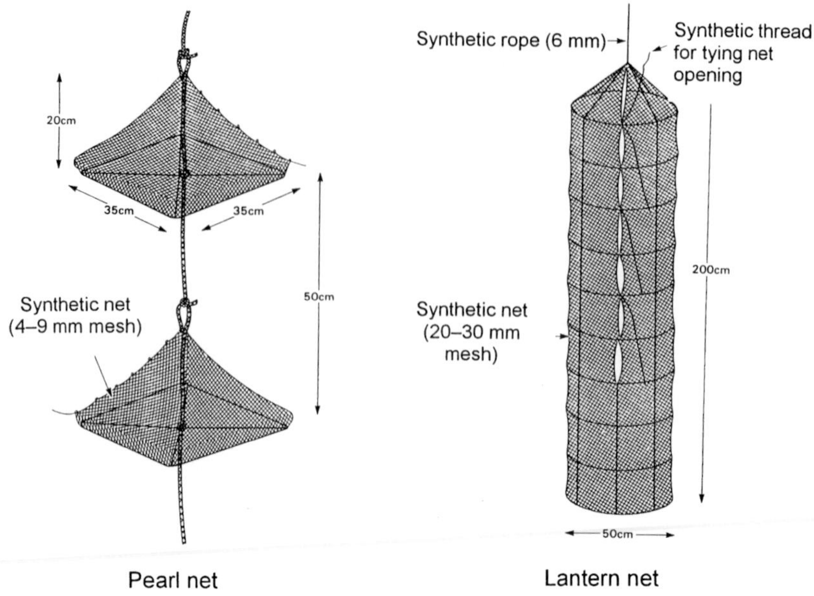

Fig. 8.3 Nets used for suspended scallop cultivation (adapted from Susumo Ito, 1991).

their inability to close their shells tightly makes them prone to desiccation and premature death. The spat *c.* 5 mm in size are graded into size groups and placed in pearl nets (Fig. 8.3) at *c.* 100 per net and suspended from long-lines (Fig. 8.4). Stocking density is adjusted within a few months to 20–50 spat per net. At the end of the first year the spat reach 50–60 mm in size with over 90% survival. Once the spat reach *c.* 30 mm, they are transferred either to lantern nets at a rate of *c.* 10 per layer or 100 per net (Fig. 8.3), or to the bottom. The lantern nets are suspended from longlines (Fig. 8.5) configured to suit local conditions. The spat are grown to market size (100 mm), usually in less than 2 years (Table 8.2).

Lantern net culture can produce a proportion of misshapen scallops. This arises from interference between neighbouring scallops, which may become interlocked ('biting' action) temporarily as their gaping shells close on each other.

In some areas, pocket nets are a popular method of cultivation. These comprise a sheet of polyethylene netting (75 cm deep × 45 cm wide × 3 cm mesh), supported by a wire frame and suspended vertically in the water column. The surface of the netting consists of a series of pockets, 2–4 across the width and up to 20 deep. Scallops are stocked at 2–3 per pocket so that the unit has *c.* 1½ times the capacity of a lantern net occupying the same

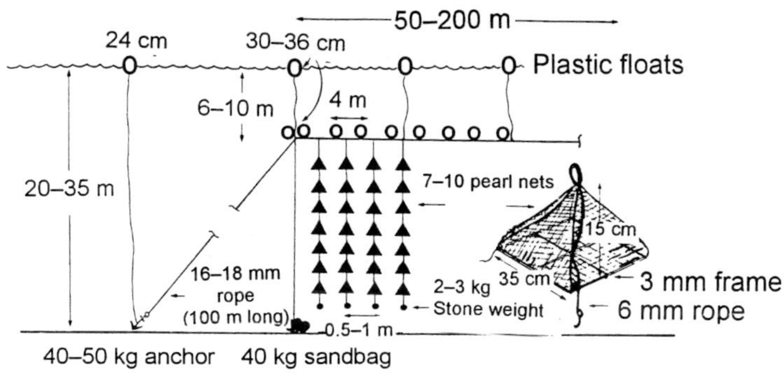

Fig. 8.4 Cultivation of small scallops in pearl nets suspended in groups of 7–10 per string. Mesh may be 3–9 mm depending on spat size (adapted from Ventilla, 1982).

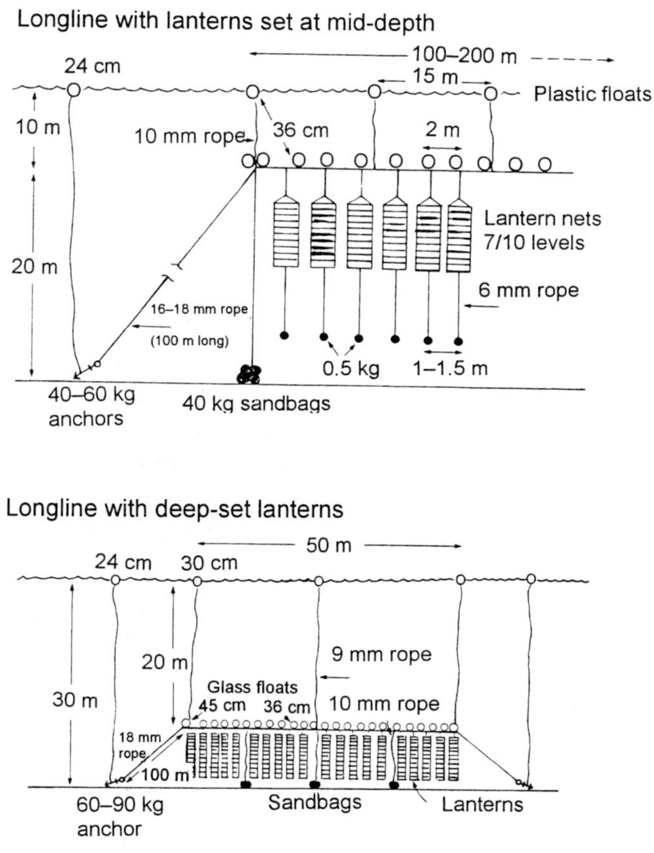

Fig. 8.5 Longlines used for suspended lantern net culture in different depths of water (adapted from Ventilla, 1982).

space. Although growth is similar to that in lantern nets, the method is said to produce scallops with both their shells convex.

Suspended culture by 'ear-hanging'

This is a popular method of cultivation used for growing scallops in shallow areas (up to 10 m deep) sheltered against inclement wave and wind action. It produces faster growth, improved survival and better shaped scallops than those grown by the other suspended net-closure methods since the scallops receive an unhindered supply of water. Although the method is initially labour-intensive and suffers from the problem of heavy fouling on the scallops, it avoids some of the high investment in equipment and the labour of servicing it.

The scallop, about 1 year old (40–60 mm in size), is attached to a dropper rope, suspended from a longline or raft, by a synthetic string or stainless steel hook threaded through a hole drilled in its 'ear' (Fig. 8.6). The scallops are spaced 8–15 cm apart, with each dropper line accommodating up to 120 shells. A 100 m longline, rigged with dropper lines 25–30 cm apart, can hold up to 45 000 shells, a capacity which is three times greater than a longline of equivalent length supporting lantern nets.

Bottom culture

Scallops grown initially in nets to *c.* 30 mm may be sown directly onto the bottom after a period of dredging to remove starfish. The method is practised mainly in Hokkaido on firm sandy soils and at depths down to 40 m. Prepared beds are stocked at a density of *c.* 5 scallops per m^2. Growth to market size (100 mm) takes 2½–3½ years with a survival of up to 30%. Higher stocking densities leads to slower growth, but in general bottom

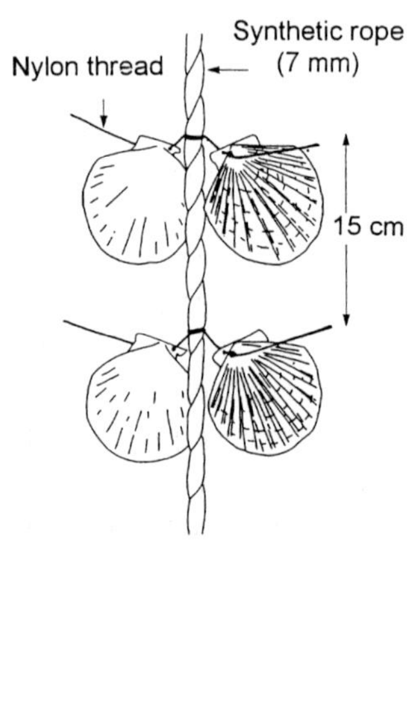

Fig. 8.6 'Ear-hanging' of scallops (adapted from Susumo Ito, 1991).

culture avoids misshapen shells common with suspended net culture. The stock is harvested by dredging with *c.* 80% of the stock recaptured.

Fouling

Fouling, as elsewhere, is one of the major constraints to scallop cultivation in Japan. A wide variety of species attach to floating equipment and to the scallops themselves. These add weight to the systems and compete for food and space. Regular husbandry, which is expensively labour-intensive and can lead to reduced growth caused by the extra disturbance, is required to control the problem. The meteoric rise in production of scallops has led to over-production and reduced growth and survival in some heavily stocked areas. The problem has been exacerbated in semi-enclosed bays and the like where poor water exchange has led to overloading of the carrying capacity of the system. Prefectural fisheries departments have introduced careful management controls by regulating the amount of scallop cultivation that takes place in these areas.

China

Of the 19 native species of scallops in China, only two, *Chlamys farreri* (Farrer's scallop) and *Mimachlamys senatoria nobilis* (the noble scallop) are cultivated. There are also two exotic species, *Patinopecten yessoensis* introduced from Japan in 1980, and *Argopecten irradians* (the bay scallop) introduced from the USA in 1982, which have become increasingly important to farmed production in China.

Scallop farming in China is a recent activity. Prior to the 1970s, wild fisheries were the sole source of scallop landings. Experimental raft culture began in the 1960s and within 20 years the exploitation of wild seed to furnish the cultivation industry had begun. In 1975 the wild fishery produced <100 t of scallops and by 1985 production had risen to 8300 t largely by cultivation of *C. farreri* (75%), *A. irradians* (20%) and *P. yessoensis* and *M. nobilis* (5%). Within another 10 years production had increased more than a hundredfold, reaching 916 000 t in 1995 and designated *P. yessoensis* in the FAO statistics (Table 8.1).

C. farreri and *P. yessoensis* are grown in the cooler waters of northern China, *A. irradians* along the central coasts, and *M. nobilis* in the warmer South China Sea. Cultivation techniques are largely similar to those developed in Japan. Seed are mainly caught in collector bags, but numerous hatcheries in the north contribute significantly to seed production, especially of *A. irradians*. Spat are cultivated, initially in pearl nets and then in lanterns or variants of these, suspended from rafts and longlines. Ear-hanging and bottom culture are both practised in some areas. The frequency of typhoons in southern waters prevents widespread use of rafts and limits production compared with northern waters.

Both *C. farreri* and *A. irradians* are relatively small scallops with short life cycles, reaching market size (50–70 mm) in 1½–2 years (Table 8.2). Specific cultivation practices have been adopted to foreshorten the cultivation cycle of *A. irradians* to less than 1 year. This has been achieved by producing spat early in the season in the hatchery. This provides sufficient time for them to grow to 5 mm, a safe size for planting in the sea, by June to take advantage of the warm summer growing period. Fast growth in the spring is further achieved by planting the spat in shrimp ponds where water temperatures are *c.* 5°C above that of the sea. In 1988, 3 billion (3×10^9) bay scallop seed were planted, yielding 50 000 t of market scallops. Thus China became the world's major bay scallop farming country within 6 years of its introduction. In 1992 landings had further risen to 130 000 t.

Nowadays the bay scallop is cultured in rotation with kelp (*Laminaria japonica*) on rafts, which utilises equipment and manpower efficiently. In recent years, polyculture of bay scallops and prawns (*Penaeus chinensis*) in shrimp ponds has been adopted. Ponds with sand bottoms and salinity above 18 psu are seeded at 10–15 scallops per m^2. Yield reaches 0.5–0.75 kg/m^2 by the end of the year. Scallops are larger and have fleshier meats than those grown on rafts and are said to improve water quality and prawn production of the ponds.

New Zealand

Japanese techniques for rearing the New Zealand scallop, *Pecten novaezelandiae* have been found to be only partially successful in New Zealand. Abundant spatfalls occurring in the Marlborough Sound/Nelson Bay area between North and South Islands have been exploited successfully by the deployment of collector bags suspended from longlines. Fouling, however, with its associated high labour and equipment costs has plagued suspended cultivation and overshadowed profitability. As a consequence, a programme of extensive cultivation has been adopted since the early 1980s, using seed collected in bags and then re-laid on the sea bed in areas of the wild fishery in the Marlborough Sound. Rates of survival in areas undisturbed by trawling reached 20% at market size of 100 mm. The problem posed by reseeding on a traditional common property resource was resolved by joint cooperation when the commercial feasibility of the programme was demonstrated.

The fishery in Marlborough Sound has shifted from a wild fishery to a managed one using seed from net bag collectors. The process is funded by a levy on the harvested scallops. The spat collectors are placed on longlines near to the reseeding area in late November–mid-December. In the 1991–92 season 300 000 bags were deployed from 142 longlines. By March, these contained an estimated 1.2 billion (1.2×10^9) seed (8–20 mm) at a rate of 1000–3000 spat per bag. These, together with re-dredged seed that had fallen from the collectors, were re-laid at *c.* 5 per m^2 onto 111 km^2 of sea bed, one of three areas used rotationally in a 3 year cycle of cultivation.

France

Scallop cultivation, although still in its infancy in France, is seen as a means of eventually contributing significantly to the dwindling wild stocks of *Pecten maximus*. Natural spatfalls, however, are unreliable and unable to provide sufficient seed for cultivation. As a consequence, a reliable hatchery and nursery technology was developed in the 1980s to provide juvenile scallops for an experimental programme of reseeding in several areas of Brittany. The aim, initially, was to build up a broodstock of adults to encourage the return of a viable fishery in the Bay of Brest. The practical aspects of the programme were successful, with 30% of the seed surviving, and achieving a broodstock of 300 t after 5 years of restocking. This programme was discontinued on the recognition that the variability of larval settlement was influenced largely by climate, with broodstock effect accounting for only a small part of the success.

Subsequently, the programme set out to grow 2 mm hatchery seed in containers on the sea bed. At 30 mm, the seed were re-laid on the bottom at a density of 10 per m^2, reaching market size of 100 mm, after 2–3 years. The overall survival rate, from larva to adult was 1%, made up of a larvae survival of 40%, cage culture survival of 35% and a recapture rate of 20–50%. Although the programme was technically successful, the production costs were approximately equal to the commercial value of the scallops, leaving no margin for profit. In 1990, the production of cultured scallops in Brittany was 30 t.

Scotland

Although commercial beds of scallops occur in many areas around the UK, Scotland has become the focus of attention for cultivation. This is because

Fig. 8.7 Cultured *Pecten maximus* at harvest. (Photo: Sea Fish Industry Authority).

of the relative success of spat collection and growth in some of the sheltered, cool, clean sea lochs of the west coast. Since the late 1960s, the scallop, *Pecten maximus* (Fig. 8.7), and queen scallop, *Chlamys opercularis*, have been extensively cultivated in sea trials conducted by the Sea Fish Industry Authority [(SFIA); formerly the White Fish Authority (WFA)]. Japanese suspended cultivation methods were applied to the local conditions, and eventually it became apparent that the low-valued queen scallop provided little scope for profit as a cultivated species.

Effort was directed to establishing the best sites for collection of seed and growing these to market size in pearl and lantern nets and by ear-hanging. Experimental deployment of spat collectors in many areas around the west coast has demonstrated the variability of set between areas and years. Settlement occurs from June to August, with a peak somewhere in between. The collectors are effective for only a short period of 2–10 days and may have to be re-deployed after cleaning if settlement is light. In some areas, spatfall comprises a mixture of queen scallop and scallop, with the former more abundant than the latter. Removal of queen scallop by handpicking is labour-intensive. It can be achieved more economically by placing a plastic cellular grid with walls *c.* 5 cm high, in the tank of mixed spat. The queen scallop spat is behaviourally more active than scallop spat and climbs upwards onto the grid, effectively separating the majority of the two populations within 24 hours. The method is affected by density and it is recommended that one spat (10 mm in size) per m^2 is used.

Commercial sets of predominantly scallop seed are collected from the sheltered waters around the Isle of Skye. Occasional failures in spatfall, however, have led to the belief that the presently small industry needs the support of hatcheries for expansion and to sustain itself in times of need. Research at the Fisheries Laboratory, Conwy, has broadly outlined the methodology required for the large-scale production of hatchery scallop seed and this is being applied by a commercial hatchery in Orkney.

Wild spat may be offered for sale outside the immediate area of collection. Transfer of seed needs to be done with care to avoid undue stress during transport, which can lead to early death. A journey time above half a day may induce stress associated with exposure to air, desiccation, temperature change, vibration and build-up of toxic wastes. Good handling, and transport facilities that keep the seed cool and moist are required to maximise survival. Basic requirements are an insulated box, cooled by a frozen coolant pack, and lined a with moisture-retaining material such as seaweed or sacking soaked with sea water. More sophisticated systems, in which the seed are drenched with water by misting, or are totally immersed in cooled water, are available to prolong the period of transport to *c.* 18 hours.

Typical growth curves of scallops, at recommended stocking densities in suspended pearl and lantern nets (Fig. 8.8) in Scottish waters indicate that the commercial size of queen scallop (60 mm) and scallop (100 mm) is reached in 3 years and 4–5 years, respectively (Fig. 8.9). Stocking density, expressed as the percentage area of a lantern net layer occupied by the scallops (% cover) is critical to growth of scallops and has been worked out in detail by SFIA (Figs 8.10, 8.11). Scallops, 25 mm in size, stocked at 10–15% cover increase in size by over 100% in the first season, whereas those stocked at 35% cover, a level considered to be overstocked, increase by 60–70% (Figs 8.10, 8.11). For practical purposes, the numbers of scallops of various sizes that can be stocked in one layer of a 50 mm diameter lantern net are shown for the suggested minimum and maximum stocking densities of 10% and 30% cover (Fig. 8.12).

As in some other countries, the high labour and equipment costs and fouling have been major problems hindering the development of suspended culture. This has directed attention to the use of the sea bed for growing juveniles from *c.* 30 mm to market size of 100 mm. Although avoiding the effect of fouling, predation from starfish, *Asterias rubens*, the edible crab, *Cancer pagurus*, and the velvet crab, *Necora puber*, poses a serious dilemma for bottom cultivation. The problem of ownership of stock re-laid on the bottom in Scotland was met in 1990 by statutory changes to the Sea Fisheries (Shellfish) Act (1967), which enabled the cultivator to secure

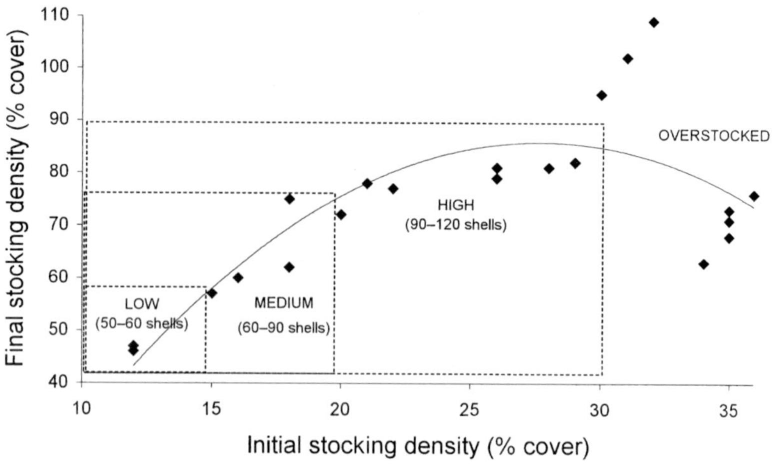

Fig. 8.11 Changes in stocking densities after first season's growth. Data for 25 mm *Pecten maximus* in 50 cm diameter lantern nets in Scotland. Trend line and number of shells per layer are shown. Data from SFIA.

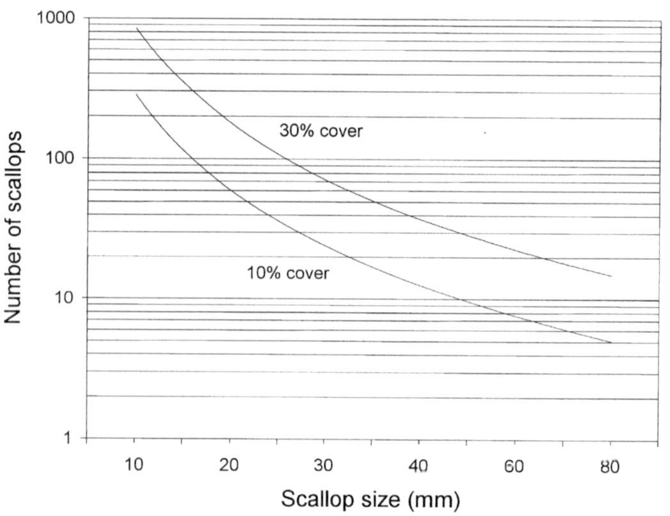

Fig. 8.12 Number of scallops per layer of a 50 cm diameter lantern net at 10% and 30% cover. Data from SFIA.

The slow uptake of scallop cultivation in other countries reflects the biological, legal and economic problems faced by the industry. The problems hindering development (Dare, 1996) are linked with various aspects of cultivation. These are:

- Hatchery: disease-related mortality of larvae (China, Chile, Australia and France).
- Wild spat: most countries experience inadequate catches due to natural variation affecting temporal and spatial abundance. Some countries have turned to hatcheries to supplement or replace wild-caught spat.
- Handling: mortality of seed during handling and transport.
- Predation: mainly from crabs and starfish in suspension systems, in on-bottom containers and on unprotected stock re-laid on the bottom.
- Bio-fouling on equipment and scallops in suspended systems.
- Dispersal of sea bed stocks: scallops that swim away from site of re-laying may be reacting to predator disturbance, overstocking and unsuitable substrate or hydrography arising from poor site selection.
- Storm losses: damage to suspended systems may reflect poor site selection.
- Low harvesting efficiency: poor recovery of crop by dredging (France).
- Phytotoxins: serious problems with PSP, DSP and ASP in Japan, Canada, USA and Scotland, where harvesting may be temporarily suspended. Also affects English, Welsh, Irish and French production in some years.
- Site security: leasing of sites and legal protection of cultivated stocks are major constraints, especially in UK, France, Spain, Italy and USA.

References and Bibliography

Bull, M. F. (1991) New Zealand scallops, *Pecten novaezelandiae*, Reeve. In: *Estuarine and Marine Bivalve Mollusk Culture* (ed. W. Menzel), pp. 319–24. CRC Press Inc., Boston.

Bull, M. F. (1994) Enhancement and management of New Zealand's southern scallop fishery. In: *Proceedings of the 9th International Pectinid Workshop, Nanaimo, British Columbia, Canada, April 1993*. Vol. 1 (eds N. F. Bourne, B. L. Bunting & L. D. Townsend), pp. 131–6. Canadian Technical Report of Fisheries and Aquatic Sciences.

Chew, K. K. & Fusui, Z. (1994) Recent developments in bay scallop, *Argopecten irradians*, culture in China. In: *Proceedings of the 9th International Pectinid Workshop, Nanaimo, British Columbia, Canada, April 1993*. Vol. 1 (eds N. F. Bourne, B. L. Bunting & L. D. Townsend), pp. 4–8. Canadian Technical Report of Fisheries and Aquatic Sciences.

Dao, J. C., Fleury, P. & Paquotte, P. (1992) *Evaluation of the Economic Cost-effectiveness of Scallop Culture*. RI-DRV-92.001-RA, Brest.

Dare, P. J. (1996) *Scallop Cultivation in the UK: a Desktop Study of R & D Progress with* Pecten maximus *During* 1975–1994. Report to the Ministry of Agriculture, Fisheries and Food, Chief Scientist's Group, London. Contract No. Ref. CSA 2967.

FAO (1999) Yearbook of Fishery Statistics for 1997. Vol. **84**. Food and Agriculture Organisation of the United Nations, Rome, Italy.

Goulletquer, P. & Héral, M. (1997) Marine molluscan trends in France: from fisheries to aquaculture. In: *The History, Present Condition, and Future of the Molluscan Fisheries of North and Central America and Europe*, Vol. **3**, Europe (eds C. L. MacKenzie, Jr, V. G. Burrell, Jr, A. Rosenfield & W. L. Roberts), pp. 137–64. US Department of Commerce, NOAA Technical Report **129**. 240pp.

Hardy, D. (1991) *Scallop farming*. Fishing News Books, Blackwell Scientific Publications Ltd, Oxford.

Millican, P. F. (1997) *The Hatchery Rearing of King Scallop* (Pecten maximus). Laboratory Leaflet CEFAS, Lowestoft.

Paul, J. D. (1991) Scallop cultivation: a simple method of separating spat of *Pecten maximus* (L.) and *Chlamys opercularis* (L.) by means of behavioural differences. *Aquaculture*, **50**, 161–7.

Susumu Ito (1991) *Patinopecten (Mizuhopecten) yessoensis* (Jay) in Japan. In: *Estuarine and Marine Bivalve Mollusk Culture.* (ed. W. Menzel), pp. 211–25. CRC Press Inc., Boston.

Ventilla, R. F. (1982) The scallop industry in Japan. In: *Advances in Marine Biology*, **20**. (eds J. H. S. Blaxter, F. S. Russell & C. M. Yonge), pp. 309–82. Academic Press, London/New York.

White Fish Authority (WFA) (1981) *An Outline of the Application of Japanese Scallop Culture Techniques in Scotland and a Review of Recent WFA Scallop Culture Research.* WFA Field Report No. 937, White Fish Authority, Edinburgh.

Yousheng Lou (1991) China. In: *Scallops: Biology, Ecology and Aquaculture.* (ed. S. E. Shumway) Developments in Aquaculture and Fisheries Science, **21**, pp. 809–24. Elsevier, Amsterdam.

Zhang, F. (1995) The Rise of the Bay Scallop Culture Industry in China. In: *The 8th International Pectinid Workshop*, Cherbourg (France), May 22–29 1991. IFREMER, Actes de Colloques, **17**.

9 Abalone cultivation

Introduction

Abalone is the common name given to the genus *Haliotis*, of which there are *c.* 20 commercial species fished significantly world-wide (Table 9.1). They are marine gastropods (along with the limpet, whelk and periwinkle, etc.) having a shallow, ear-shaped shell with a series of respiratory holes along the dorso-lateral shell margin (Fig. 9.1). In some species, the inside of the shell is brilliantly iridescent with a lining of mother-of-pearl, making it an attractive material for the manufacture of jewelry items, and for the production of blister pearls. Abalones occupy the low intertidal and high subtidal zones of exposed coasts in clear, well-oxygenated, high salinity sea water. They attach to the underside of rocks and boulders with their powerful muscular foot, from which they are not easily dislodged. They feed on drift seaweeds (macroalgae), which they trap with their foot and consume by rasping away the layers with teeth-like radulae. Their feeding mechanism is, therefore, distinctly different from that of the filter-feeding bivalves.

Although abalones occupy a lowly position in world landings statistics (12 900 t in 1997, Table 9.2), comprising <1% of the total molluscan landings, they are highly prized gastronomically in many countries, especially in east Asia. They command extraordinarily high prices of around US $22 000 per tonne live weight and three times this value in Japan (1993 prices). Japan and China are major world consumers (importing from USA, Hong Kong and other countries), and together with SE Asia, consume more than 80% of the world total. Although FAO statistics list *H. gigantea* as the principal landed species in Japan (Table 9.2), it contributes only *c.* 14% of the total, with *H. discus hannai* (46%), *H. discus* (26%) and *H. sieboldii* (14%) making up the rest.

The relative scarcity and high prices on world markets of abalones has created much interest in their cultivation, with many countries undertaking R&D in support of their emerging new industries. The Japanese, following the collapse of their wild fisheries after World War II, pioneered abalone farming. Hatcheries were used to produce 15–20 mm juveniles for reseeding natural areas. This strategy is still used today with a small number of farms growing abalones to market size in land-based tanks and from rafts.

Table 9.1 Commercially important species of *Haliotis* [adapted from Jarayabhand & Paphavasit, 1996 (after Hahn, 1989 and Fallu, 1991)]

Species	Common name	Location	Shell length (mm)
H. rufescens★	Red abalone	Mexico, Pacific coast of USA	>275
H. fulgens★	Green, southern green or blue	" "	125–200
H. corrugata	Pink or corrugated	" "	150–175
H. sorenseni	White or sorensen	Pacific coast of USA	125–200
H. assimilis	Threaded	Mexico, Pacific coast of USA	>200
H. cracherodii	Black	Pacific coast of USA	75–125
H. walallensis	Flat or northern green	Pacific coast of Canada and USA	75–125
H. kamtschatkana	Northern or pinto	Pacific coast of Canada and USA	100
H. discus hannai★	Ezo awabi	North Japan, North China, Siberia	180–200
H. discus	Kuro awabi (black)	" "	200
H. gigantea	Madaka	South Japan	250
H. sieboldii	Megae	South Japan	170
H. diversicolor diversicolor#	Small	Hong Kong	
H. diversicolor supertexa#	Tokobushi, small abalone	South Japan, Taiwan	50
H. asinina#	Mimigai, donkey's ear	South Japan, Thailand	70–100
H. rubra★	Blacklip	Australia	120–140
H. laevagata	Greenlip	" "	130–140
H. roei	Roe's	" "	70–80
H. iris★	Paua or black	New Zealand	170
H. australis	Silver or queen paua	" "	125
H. virginea	Virgin or white paua	" "	70
H. tuberculata★	Ormer	Europe (France and Channel Isles)	120
H. midae★	Perlemoen	South Africa	90

★ farmed species; # tropical species

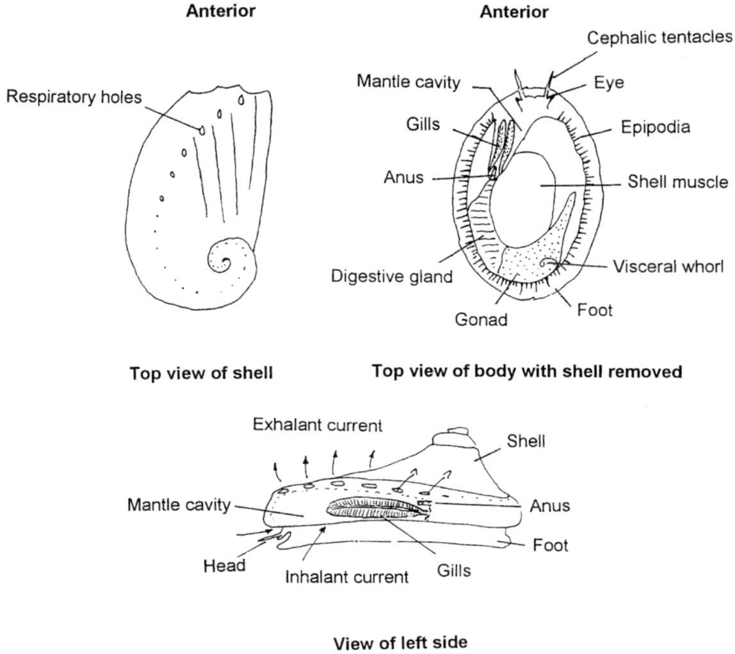

Fig. 9.1 Diagrammatic views of abalone showing layout of some organs and flow of water into and out of mantle cavity.

Table 9.2 Main areas of abalone production (FAO, 1999)

Species	Common name	Country	1987	1997	10 year average*
			Thousands of tonnes		
Haliotis gigantea	Giant abalone	Japan	4.2	2.2	2.9 (76%)
Haliotis midae	Perlemoen abalone	South Africa	0.7	0.3	0.6 (53%)
Haliotis rubra	Blacklip abalone	Mostly Australia	6.7	5.2	5.5 (94%)
Haliotis spp.	Various	Taiwan	1.1	2.2	1.3 (174%)
		Mexico	1.5	0.9	2.1 (43%)
		New Zealand	1.3	1.2	1.2 (98%)
		World total all species	16.5	12.9	14.8 (88%)

* 1997 values as % of 10 year average (1987–96)

Hatchery, nursery and on-growing techniques are now well established world-wide and provide the opportunity for the shellfish industry to contribute significantly to world production. Commercial abalone farms in Japan, China, Korea, South Africa, USA, Australia, New Zealand, Mexico,

France and Channel Isles and Ireland, however, as yet contribute probably no more than *c.* 5% of world landings. Most commercial species of abalones are from temperate or semi-tropical waters and are slow growing, reaching a relatively large size (>150 mm, Table 9.1) in their natural habitats after up to 10 years or more. The cultivation industry, however, seeks to market fairly small, "cocktail" abalones (*c.* 50–80 mm) that can be grown relatively quickly within a period of 3–4 years.

Cultivation procedures

Various countries have adopted a variety of cultivation strategies, but all include initially the production of juveniles via hatchery and nursery facilities. Once the juveniles have reached a suitable size of *c.* 10 mm, they may be transferred to outdoor tanks for growing to market size (50–80 mm) or to an intermediate size of *c.* 30 mm. They are then transferred to sea cages or barrels, suspended from a longline or raft, for growing to market size. Adapted lobster pots strung along the sea bed have also been used for on-growing in Ireland. The reseeding of natural areas or artificial reefs as a conservation measure for recreational fishing or for ranching, is a strategy used or tried in Japan, China, USA and Australia.

Hatchery culture

The technology required is relatively straightforward. It involves conditioning and then spawning broodstock to produce eggs and sperm, nurturing the veliger larvae through a short pelagic phase lasting for about 5 days, and then, following metamorphosis of the larvae, caring for the spat until sufficiently large (3–5 mm) for weaning on to a macroalgal diet. The larvae are sustained by stored energy reserves and require no external source of feeding.

Broodstock conditioning

Adult abalones are kept in tanks supplied with running sea water at near ambient summer temperatures (10–20°C for temperate species), a salinity of 30–35 psu and low suspended particle content. The length of conditioning varies with abalone species and proximity to natural spawning at initial stocking. It is around 1000–1500 day-degrees [number of days × temperature above biological zero (*c.* 7°C for temperate species)] and may be as high as 3500 day-degrees for *H. gigantea*. A light/dark regime may be used. This is because abalones are negatively phototactic, moving to the bottom of an illuminated tank or moving freely around the tank in darkness. In New

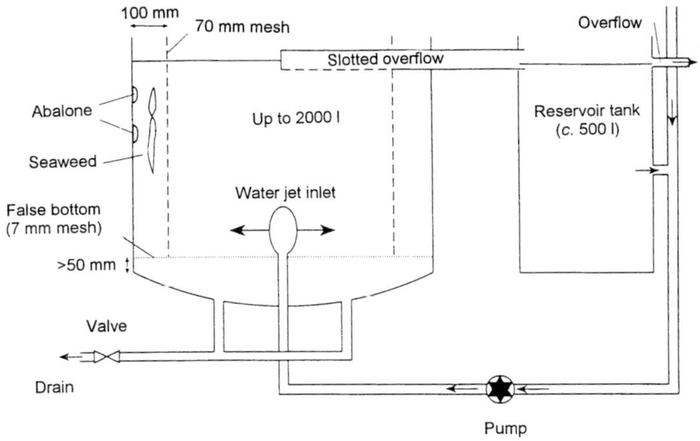

Fig. 9.2 Diagram of broodstock holding tank (adapted from Tong *et al.*, 1992).

Zealand (*H. iris*), north Japan (*H. discus hannai*) and Europe (*H. tuberculata*), the regime is 12 hours dark and 12 hours light, in south Japan (*H. discus hannai*) it is 16:8, and in California (*H. rufescens*) 24 hours dark. The broodstock are fed on macroalgae, usually collected locally. The species of alga depend on their suitability as food and on availability. Generally, red seaweeds are better than brown, while green seaweeds have little food value. In Japan, *H. discus hannai* is fed *Macrocystis pyrifera* and *Undaria pinnatifida*, fresh in season or frozen/dried out of season with no apparent loss in food value. In Wellington, New Zealand, locally common brown seaweeds, such as *M. pyrifera, Lessonia variagata* and *Durvillaea antartica*, are added to the tanks twice weekly in summer and weekly in winter.

The quantity of food added is calculated on the feeding rate of abalones, which is 10% of the body weight in summer and 4% in winter. This equates to 12–16 kg per feed for 100 abalones of 100 mm shell length. Tank design may be simple or specially constructed to facilitate ease of feeding by concentrating the macroalgae within an inner wall of plastic mesh, in close proximity to the broodstock (Fig. 9.2). This tank system, accommodating 80–180 abalones per 1000 l capacity, is supplied with sea water at *c*. 500 l per hour. The broodstock attached to the walls, receive turbulent, well-oxygenated water via a centrally placed water jet. Faeces and uneaten food, falling to the bottom of the tank, are flushed out weekly.

Spawning

When ripe the abalones, whose sexes are separate, may spawn naturally or may be induced to spawn. Eggs and sperm are released into the mantle cavity and then flushed to the outside with the exhalant current through the respiratory holes in the shell, where under natural conditions fertilisation occurs. To ensure a degree of control over fertilisation, the sexes are spawned separately in small tanks (10–15 l). Induced spawning may be achieved by cycling the temperature of the sea water *c.* 5°C above and below ambient (15–18°C) or by subjecting the broodstock to natural cues with the addition of small quantities of eggs or sperm. Other methods include air drying, using sea water irradiated with UV light, or chemicals such as a mixture of hydrogen peroxide and sodium hydroxide. UV light and hydrogen peroxide are the most successful, but the preferred method is a matter of choice, judged by the operator on their success rate with the species with which they are working.

Larvae

Fertilisation is achieved by adding sperm at *c.* 25 000 per litre to a suspension of eggs. After half an hour the eggs are washed free of sperm, using a sieve (*c.* 60 µm mesh), and placed in a hatching tank where they sink to the bottom (Fig. 9.3). Egg development, occurring initially within the egg membrane, is fairly rapid (Fig. 9.4). The trochophore stage is reached within *c.* 24 hours and on hatching from the egg membrane swims to the water

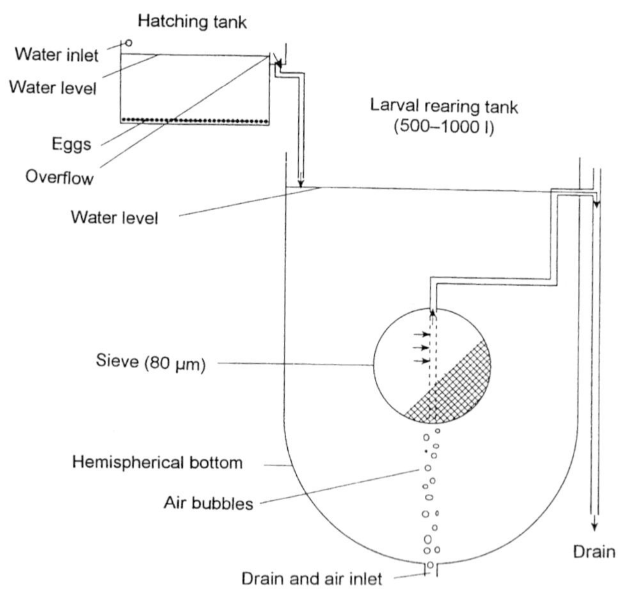

Fig. 9.3 Diagram of a hatching tank and larval rearing tank fitted with a banjo sieve (from Tong *et al.*, 1992).

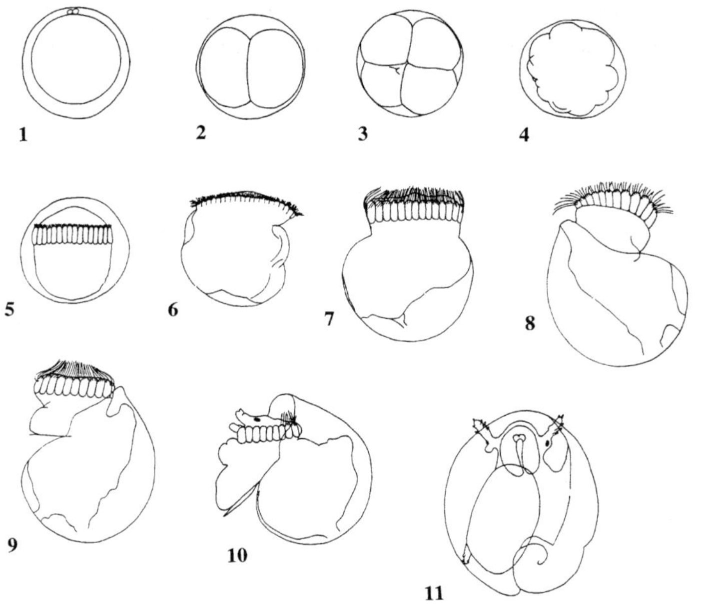

Fig. 9.4 Development of abalone egg to post-larva stage (from Tong, 1982). (1) Fertilised egg (230 μm) with polar bodies. (2) 2-cells. (3) 4-cells. (4) 12-cells. (5) Trochophore before hatching. (6) Hatched trochophore with developing shell. (7) Early veliger larva. (8) Veliger undergoing torsion. (9) Veliger after torsion, with foot and operculum. (10) Veliger with early tentacles and eyespots. (11) Post-larva with developing tentacles and mouth parts.

surface. The flow of water in the system developed in New Zealand (Fig. 9.3) flushes the trochophores from the hatching tank into a larval rearing tank supplied with sea water at *c.* 41 per minute. The larvae, stocked at *c.* 6 per litre, develop into the veliger form, undergoing torsion (twisting of the body relative to the shell, which is unique to some groups of gastropods) and further development of the foot, operculum, mouthparts, tentacles etc. After *c.* 7 days at 15°C, the larva metamorphoses into a bottom-living spat equipped with functioning mouthparts, sensory organs and foot to allow it to seek its food.

Spat

Spatfall is enhanced primarily by the provision of settlement plates coated with benthic diatoms and other microalgae and bacteria, of an unspecified nature. Commonly identified diatoms include species such as *Gramatophora*, *Melosia*, *Nitzschia*, *Navicula* and *Cocconeis*. The coating is prepared by holding the plates for about a month in outdoor tanks supplied with running sea water. The coating process generally is rather hit-and-miss, producing diatoms etc. with good, poor or indifferent food value. Diatom quality and abundance may be improved by using cultured diatoms, or by fertilising the water with nitrates, phosphates and silicates.

Control of the algae film on the settlement plates is critical for consistent survival. Frequently, high spat mortality within 4 weeks of settlement is food- or density-related, caused by the wrong combination of diatoms or too many spat on the plates. Extra food or more space may be provided by adding more algae-coated plates to the tanks, or if the coatings become too dense or too thin by controlling the shading over the tanks. In Japan, the plates may also be pre-conditioned with the mucous trails of adult abalones, and in USA and Mexico a settlement stimulant, GABA (gamma-amino-butyric acid) may be used.

The settlement plates are plastic sheets held vertically in tanks. They may be flat or corrugated, matt-surfaced for easy adherence of the diatoms and spat, and translucent for improved light penetration. A system developed in New Zealand uses plates (50 cm wide × 20 cm deep) suspended 10 cm apart in a V-shaped tank, giving a high surface area-to-volume ratio.

Nursery

Once settled, the spat are kept on the feeder plates in indoor tanks supplied with running sea water at *c.* 15°C and illuminated, usually with natural light to sustain the microalgae coating. After 4–6 months, the spat (*c.* 5 mm) are weaned onto a macroalgae diet, the species depending on availability. The red seaweed, *Gracilaria* is a good quality food, and in New Zealand is added to the V-shaped tanks at a rate of 50 g every 4–5 days. Good quality artificial diets, in the form of pellets, are also available and may be fed to spat from *c.* 3 mm in size.

Nursery systems for growing the spat to 10–30 mm for sea culture or to market size of 50–80 mm are usually outdoor tanks or raceways. With temperate species, winter temperatures must exceed a critical minimum limit of *c.* 6°C to avoid undue mortality and to ensure that growth is continuous. A design used in California consists of a series of concrete raceways (3 × 2 × 1 m deep) subdivided with compartments reaching from below surface water level to within 20 cm of the bottom. Sand-filtered sea water flowing in at one end at 3 tank volumes per hour, is directed along the bottom of the raceway and into each compartment by vigorous aeration or by air-lift pump. Each raceway was stocked with 5000 20 mm abalones. Other systems of similar basic design may use a plastic mesh cage (e.g. 150 × 70 × 33 cm; 3 mm mesh) in each compartment, with inverted plastic gutters providing shelter for the abalones. Stocking density depends on abalone size, ranging from 5000 per cage at 8–10 mm to 1000 at 20–26 mm. The abalones are fed on a mixed species diet of macroalgae (Table 9.1 provides an indication of the geographical range of species) cut up into small pieces, or pellets.

Handling

Removal of the juveniles from the tanks (or sea containers) for servicing, grading, density adjustments, marketing etc., is a necessary, intermittent husbandry activity. However, because abalones attach so strongly to surfaces, removing them by hand can cause tissue damage, stress and eventually high mortalities. The problem stems from their inability to repair tissue damage quickly, by the absence of a blood coagulating mechanism. Handling is made safer and more effective by anaesthetising the abalones. Anaesthesia may be achieved by a variety of chemicals. Two found to be effective with *H. midae* are magnesium sulphate ($MgSO_4$) and 2-phenoxyethanol, when used in sea water at a salinity of 35 psu and temperature of 18°C. The concentrations used for animals 5–15 mm, 20–50 mm and 60–90 mm in size, were 4, 14, and 22 g respectively per 100 ml of sea water for $MgSO_4$, and 0.05, 0.2 and 0.3 ml per 100 ml of sea water for 2-phenoxyethanol. The response time for complete anaesthesia increased with size, from <1 minute to <10 minutes, with no mortalities occurring when the animals were returned to fresh sea water.

In New Zealand, *H. iris* spat are removed from the feeder plates by anaesthetising with benzocaine (ethyl p-aminobenzoic acid) dissolved in 95% ethyl alcohol. The plates are immersed for 5 minutes in sea water (15°C) containing benzocaine at a final dilution of 100 ppm. After treatment, the juveniles are gently hosed from the plates, graded as required and revived in running sea water.

Other chemicals used elsewhere are a 0.5% solution of chloral hydrate (90% fall off after 3 minutes; 15 minutes recovery time), and a 20% solution sodium diethylbarbiturate (100% fall off in 30 minutes; 150 minute recovery time).

Macroalgae diet

Despite the variation in food value of macroalgae between species and between seasons, abalones are fed those that are locally abundant (Table 9.3 and Table 9.5). The attractiveness of an alga as a food to abalones may be due to its texture (ease of ingestion) or nutritional value, coupled with the presence of chemical feeding stimulants (e.g. certain amino acids, fatty acids) or repellants (e.g. phenolic compounds in brown seaweeds). Abalones generally prefer red rather than brown seaweeds, with green seaweeds having poor food value. Japanese abalones, however, prefer brown seaweeds, although trials with *H. discus hannai* in Ireland showed that the order of preference

Table 9.3 Calculated annual food requirements of 100 000 *Haliotis tuberculata*, reaching market size (60 mm; 35 g) in 4 years (from La Touche *et al*., 1993)

Year	Total weight of ormers (kg)	Wet weight of *P. palmata* (tonnes)
1	2	–
2	300	2.1
3	1500	10.5
4	3500	24.5

was *Palmeria palmata* (red, dulse) = *Alaria esculenta* (brown) > *Laminaria digitata* (brown kelp or oar weed) = *L. saccharina* (brown sugar kelp) = *Ulva lactuca* (green, sea lettuce). Similar trials with *H. tuberculata* showed that the preference was *P. palmata* > *A. esculenta* = *U. lactuca* = *L. digitata* > *L. saccharina* = *Chondrus crispus* (red, Irish moss or carrageen). In New Zealand, trials with *H. iris* revealed that the order of preference for seven macroalgae was *Gracilaria* > *Glossophora* > *M. pyrifera* > *Lessonia variagata* > *Chamia* > *U. lactuca* > *Pterocladia*, with *Gracilaria* producing twice the growth rate of *M. pyrifera*.

The feeding rate of abalone depends on a number of factors including age, species and sea water temperature. Adults in conditioning tanks are usually fed at *c*. 10% of their body weight (on live weight basis) per day. Actively growing juveniles may require feeding at up to 20–40% of their body weight per day. An average annual conversion rate of 7:1 (7 g of live alga per g of abalone live weight) may be used to calculate the food requirements of a farm (Table 9.3). In New Zealand, a food requirement of 1400 tonnes per year of algae has been calculated for a farm producing one million *H. iris* per annum.

The problems associated with providing large amounts of live algae to sustain an abalone farm have led to the search for alternative supplies. In a number of countries macroalgae, cultivated either as a human food or for alginate production, may be used. Elsewhere, macroalgae may be used as biological filters for nutrient stripping of water prior to its return to the sea (e.g. fish farms in USA, Israel; shrimp farms in SE Asia). These can provide an opportunity for polyculture on a small scale, by incorporating bivalves and abalones for feeding on the micro- and macroalgae produced in the systems. In Korea a number of land plants, especially Chinese cabbage and mulberry, have been found to be suitable as foods when the seaweeds normally fed are in short supply.

Artificial diets

These offer most promise as a replacement or supplement to a macroalgae diet in tank systems, overcoming the problems of availability, abundance, collection, storage, culturing and cost. At the same time, an artificial diet offers an opportunity for the manufacturer to formulate a diet that yields optimum growth and survival at low cost. The search for such a diet began in Japan in the 1960s. Commercial production is now under way in Japan and China while research and development is being undertaken in many other countries with an interest in abalone farming (e.g. South Africa, USA, Korea, Mexico, Canada, France, Australia and New Zealand).

Artificial diets are a nutritionally balanced mixture of a variety of substances, mostly unavailable naturally to abalones, formulated largely to promote good growth and survival. They are usually manufactured as neutrally buoyant, dry pellets prepared in disc, ribbon and string and even frond shape, mimicking an alga thallus. In addition to suitable presentation, the pellet must retain its integrity in sea water and have a low leaching rate of constituent components for a sufficiently long time for it to be eaten. It must also be readily available to the abalones and not create problems of waste disposal and pollution within the system. Formulations between diets vary enormously. A survey of commercial and experimental diets showed a range of gross compositions (Table 9.4) with precise information on some components, usually particle size, binders and attractants, subject to confidentiality.

The low metabolic rate of abalones means that their energy requirement is low. The maintenance ration of a 25 g *H. rubra* at 18°C has been estimated

Table 9.4 Composition of artificial diets (from Flemming *et al*., 1996)

Major components	Content (% dry weight)	
Protein	20–53	Mainly from fish meal, casein or soya bean
Synthetic amino acid additives	0.2–0.5	Including lysine, methionine, threonine and arginine
Total lipid	1.5–5	Variety of fish and/or vegetable oils with essential fatty acids (n-3 at *c*. 20% and n-6 at *c*. 3%)
Total carbohydrate	32–60	Vegetable meal/flour (e.g. soya, maize, rice starch)
Binders	2–30	Alginates, gelatin, starch, gluten, bran, semolina, maize, whey, crude fibre, etc.)
Vitamins	2.7–7.5	E.g. vitamins A, C, D and E
Minerals	1–3	Calcium and potassium
Stimulants/attractants	Up to *c*. 20%	Kelp meal, red algae, spices and herbs
Antioxidants	–	Vitamin E

Table 9.5 Growth, survival and food of juvenile *Haliotis* in barrel (or cage) culture (adapted from Aviles *et al.*, 1996)

Species	Annual growth rate (mm per month[#])	Annual survival (%)	Location (T°C range)	Algae diet	Stocking density (no./l)	Barrel (cage) volume (l)
H. tuberculata	0.8	94	West Ireland (6–16)	*Palmeria palmata, Chondrus crispus, Ulva lactuca*	2.0	250
H. iris	0.8	70–80	Wellington, New Zealand (n.d.)	*Macrocystis pyrifera*	2.2	134
H. discus hannai	1.9	n.d.	N. China (5–25)	*Laminaria japonica*	0.8	100 (cage)
H. fulgens	2.1–2.4	88	Baja California, Mexico (15–23)	*M. pyrifera* or *Egregia menziesii*	1.3	95 & 190
H. corrugata	0.3	74	″	″	2.5	95
H. rufescens	2.2 (low density) 1.5 (high density)	66 39	″ ″	″ ″	1.3 2.5	95 95
H. rufescens	1.9	96	California, USA (n.d.)	*Neriocystis luetkeana*	10, then 2.5 later	200

H. rufescens	1.2 (barrel)	76	Baja California, Mexico (14–19)	*M. pyrifera*	7.5 initially	200
	1.7 (cage)	83	″	*E. menziesii*	12 initially	1000 (cage)
H. rufescens	2.2	80	Baja California, Mexico (16–25)	*M. pyrifera, E. menziesii*	1.8	100
H. fulgens	3.2	~100	Baja California, Mexico (16–24)	*M. pyrifera, Eisenia arborea, Ulva* spp., *Laminaria* spp., *Enteromorpha* spp.	n.d.	200
H. fulgens; H. rufescens	1.6–2.0	78 (at 10 mm); 93 (at 35 mm)	California, USA (8–21)	*M. pyrifera, E. menziesii, Gracilaria* spp.	initially 12.5, then 2.5 later	200
H. fulgens	1.5	86	Baja California, Mexico (12–23)	*M. pyrifera, E. arborea*	1.2	100
H. rubra	1.8	n.d.	Tasmania, Australia (10–18)	*M. pyrifera*	n.d.	n.d.

n.d. no data; # calculated from μm/day in original Table

to be *c.* 0.2–0.3 kJ per day and when feeding on *Jeannerettia lobata*, 1.2 kJ per day. On average, an artificial food might contain *c.* 15 kJ per g, of which a 25 g abalone might consume *c.* 0.5 g per day at a daily intake of 2% body weight. Thus artificial foods generally contain energy surplus to requirements and it has been suggested that this is converted to glycogen stored in the foot, adding enhanced flavour to the marketed meat.

Abalone farmers require average growth rates of 2–3 mm per month to ensure that the juveniles reach 50 mm in 2 years and market size of 70–80 mm in 3 years. Although information is lacking on performance with long-term trials, formulated diets appear to be moving in the right direction with some meeting the growth requirements in short-term trials. Despite the evidence of the experimentalists, it is a fact that some farms world-wide use artificial diets, either as a supplement or replacement for macroalgae, in rearing their abalones during some part of the cultivation cycle.

Sea culture

Juvenile abalones may be cultured in sheltered areas of the sea in containers usually suspended from a longline, raft or, in some cases, a fixed structure. The site must have a good flow of clean sea water with low content of suspended matter and must provide easy access to allow for regular feeding and maintenance. The containers may be converted or purpose-made barrels, with end or side panels replaced with plastic mesh to allow ample exchange of sea water, or cages of mesh. The containers are fitted with secure access holes for regular feeding, and have solid or mesh inner walls or floors to provide a larger surface area (up to *c.* 3 m^2) for the habitation of the abalones. Barrel size is about 1 m × 0.8 m diameter, with a volume of up to 250 l; cages may be larger (Table 9.5).

One commercial company in California, believing that barrel culture is uneconomic, has developed a cage culture system. The cages are filled with vertical plastic sheets, 20 cm apart, and suspended from a raft. Vigorous aeration from *c.* 15 cm beneath the cages ensures that well-oxygenated water reaches, and removes wastes from the innermost levels of the sheets. The cages, stocked with 20 000 juvenile *H. rufescens* (10 × greater than a conventional 250 l barrel) are said to achieve >25% better growth than the conventional barrel. By reducing the number of containers and therefore the labour costs associated with feeding and maintenance for a given level of output, production costs have been reduced by half.

Stocking densities must be reduced, and mesh apertures of the panels increased to ensure good exchange of water with size of animal. Details for *H. tuberculata* are shown in Table 9.6. The grazing activity of the abalones

Table 9.6 Stocking density of *Haliotis tuberculata* in barrels and mesh size of ventilation panels (from La Touche *et al.*, 1993)

Shell length (mm)	No./l of container volume	Mesh size (mm)
8–10	20	2–3
10–15	20	4
15–26	20	6
26–41	5	10
41–56	2	15
56–70	1	15

over the inside walls and mesh panels provides a self-cleaning service so that a minimum number should be stocked to prevent meshes from becoming fouled with newly settled seaweed.

Growth and survival

The performance of abalones in culture depends on species, environment, feeding and handling etc. The following three examples provide some information on growth and survival with different species in different countries. In Taiwan the tropical small abalone, *H. diversicolor supertexa*, is cultivated mainly in concrete ponds. Traditionally these were built intertidally and flushed daily with tidal incursions. Nowadays new ponds provided with pumped sea water are built on shore owing to Government restrictions on the use of the foreshore. Hatchery-reared seed (3–10 mm), planted at a density of 3500–7000 per m^2 on settlement plates, are fed chopped *Gracilaria* and *Ulva*. On reaching 15 mm the juveniles are moved to grow-out ponds with a layer of stones on the bottom for shelter. The abalones, stocked initially at 150–250 per m^2, reach market size of 60 mm in 4 months with a survival of *c.* 70% and yield of *c.* 4 kg/m^2 (20–30 g mean weight) per annum. Growth is seasonal and fastest during April–September (Fig. 9.5).

In Korea, on-growing of abalones is done in sea cages, typically taking 4 years for juveniles *c.* 30–40 mm to reach market size of >60 mm (Fig. 9.6), with >85% survival.

In Ireland, growth of *H. tuberculata* may be strongly influenced by low winter temperatures, stopping completely below 7°C (Fig. 9.7). In these circumstances, it is important to select an on-growing site that avoids such restrictive growth.

Predators

The natural predators of abalone include octopus, crab and starfish. Only the last two, if uncontrolled, may present problems in nursery culture

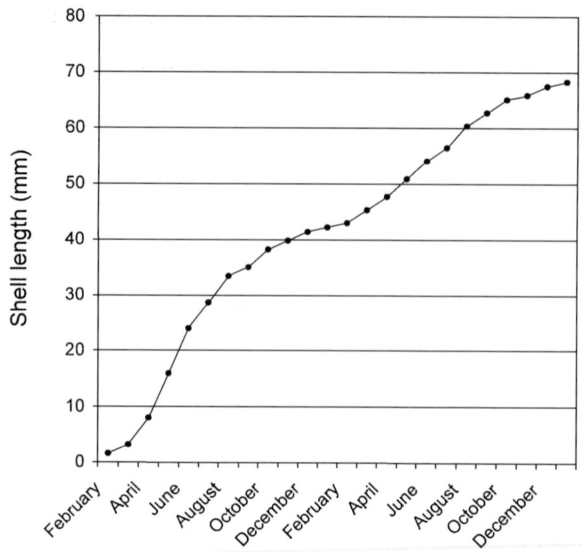

Fig. 9.5 Growth of *Haliotis diversicolor supertexa* in ponds (redrawn from Chen, in Hahn, 1989).

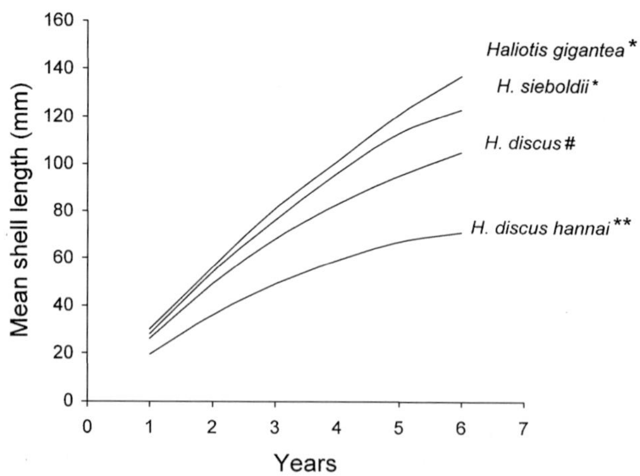

Fig. 9.6 Growth of four temperate species of *Haliotis* in sea containers at three sites in Korea; ⋆at Jeju, 14–26°C; # Ilkwang, 11–22°C; ⋆⋆Ullungodo, 8–24°C (redrawn from Sung, in Hahn, 1989).

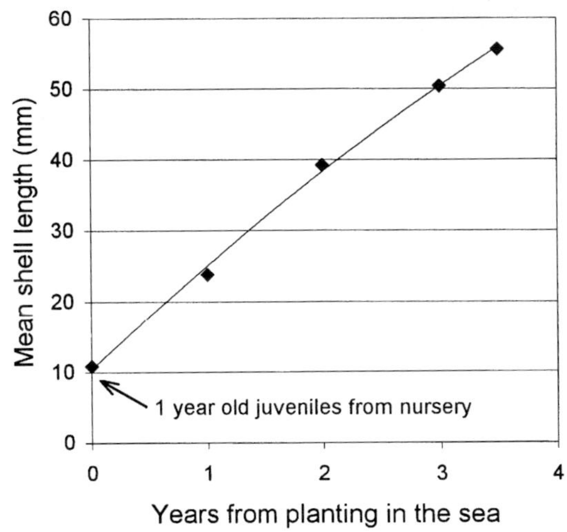

Fig. 9.7 Growth of *Haliotis tuberculata* in sea cages in western Ireland (from data in La Touche *et al.*, 1993).

in tanks or sea cages, gaining access via added macroalgae or as planktonic larvae.

Pearls

A small number of commercial abalone farms world-wide (Japan, Korea, New Zealand, USA, Canada) produce valuable blister pearls in addition to meats. The technique involves the insertion of a plastic nucleus ('mabe') between the shell and mantle. Over a period of *c.* 3–4 years the foreign body is enclosed in nacre to form a hemispherical or half-globe 'mabe pearl' with a size of 10–20 mm. High quality pearl production requires stringent husbandry and management controls, with an added risk of higher mortality due to the extra handling involved.

Prospects for the future

Dwindling natural fisheries and high demand and prices on the world market make the cultivation of abalones an exciting prospect. Hatchery, nursery and on-growing technologies are fairly well researched, and are capable of producing cocktail-sized animals within a period of up to 3–4 years depending on species. The continuous provision of macroalgae as food for juveniles and adults may present problems in some areas. The production of good quality artificial diets is seen as a means of subverting this problem provided the cost is kept sufficiently low to maintain profitability of abalone culture.

References and bibliography

Anon. (1995) Abalone grow best in sea cages. *Fish Farming International*, **22** (1), 10–11.

Aviles, J. G. G. & Shepherd, S. A. (1996) Growth and survival of the blue abalone *Haliotis fulgens* in barrels at Cedros Island, Baja California, with a review of abalone barrel culture. *Aquaculture*, **140**, 169–76.

Chen, H.-C. (1989) Farming the small abalone, *Haliotis diversicolor supertexa*, in Taiwan. In: *Handbook of Culture of Abalone and Other Marine Gastropods*. (ed. K. O. Hahn), pp. 265–83. CRC Press, Florida, USA.

Fallu, N. (1991) *Abalone Farming*. Fishing News Books, Blackwell Science Ltd, Oxford.

FAO (1999) Yearbook of Fishery Statistics for 1997. Vol. **84**. Food and Agriculture Organization of the United Nations, Rome, Italy.

Flemming, A. E., Van Barneveld, R. J. & Hone, P. W. (1996) The development of artificial diets for abalone: a review and future directions. *Aquaculture*, 140, 5–53.

Hahn, K. O. (1989) ed. *Handbook of Culture of Abalone and Other Marine Gastropods*. CRC press, Florida, USA.

Jarayabhand, P. & Paphavasit, N. (1996) A review of the culture of tropical abalone with special reference to Thailand. *Aquaculture*, **140**, 159–68.

La Touche, R., Moyan K. & Twomey, W. (1993) *Abalone On-growing Manual*. Aquaculture Explained, **14** Bord Iascaigh Mhara, Dublin.

Mai, K., Mercer, J. P. & Donlan, J. (1993) Comparative studies on the nutrition of two species of abalone, *Haliotis tuberculata* L. and *Haliotis discus hannai* Ino. I. Effects of algal diets on growth and biochemical composition. *Invertebrate Reproduction and Development*, **23**, 75–88.

Mai, K., Mercer, J. P. & Donlan, J. (1994) Comparative studies of two species of abalone, *Haliotis tuberculata* L. and *Haliotis discus hannai* Ino. II. Amino acid composition of abalone and six species of macroalgae with an assessment of their nutritional value. *Aquaculture*, **128**, 115–30.

Mai, K., Mercer, J. P. & Donlan, J. (1995a) Comparative studies on the nutrition of two species of abalone, *Haliotis tuberculata* L. and *Haliotis discus hannai* Ino. III. Response of abalone to various levels of dietary lipid. *Aquaculture*, **134**, 65–80.

Mai, K., Mercer, J. P. & Donlan, J. (1995b) Comparative studies on the nutrition of two species of abalone, *Haliotis tuberculata* L. and *Haliotis discus hannai* Ino. IV. Optimum dietary protein level for growth. *Aquaculture*, **136**, 165–80.

Mai, K., Mercer, J. P. & Donlan, J. (1996) Comparative studies on the nutrition of two species of abalone, *Haliotis tuberculata* L. and *Haliotis discus hannai* Ino. V. The role of polyunsaturated fatty acids of macroalgae in abalone nutrition. *Aquaculture*, **139**, 77–89.

Oakes, F. R. & Ponte, R. D. (1996) The abalone market: opportunities for cultured abalone. *Aquaculture*, **140**, 187–95.

Shepherd, S. A., Tegler, M. I. & Guzman del Proo, S. A. (eds) (1992) *Abalone of the World, Biology, Fisheries and Culture*. Fishing News Books, Blackwell Science Ltd, Oxford.

Sung, K. Y. (1989) Abalone farming in Korea. In: *Handbook of Culture of Abalone and Other Marine Gastropods*. (ed. K. O. Hahn), pp. 255–63. CRC Press, Florida, USA.

Tong, L. J. (1982) The potential of aquaculture of paua in New Zealand. In: *Proceedings of the Paua Fishery Workshop*. (eds J. M. Akroyd, T. E. Murray & J. L. Taylor), pp. 36–40. Fisheries Research Division Occasional Publication, **41**, Wellington, New Zealand.

Tong, L. J., Moss, G. A., Redfearn, P. & Illingworth, J. (1992) A Manual of Techniques for Culturing Paua, *Haliotis iris*, through the Early Juvenile Stage. New Zealand Fisheries Technical Report, **31**, Fisheries Research Division, Wellington, New Zealand.

White, H. I., Hecht, T. & Potgieter, B. (1996) The effect of four anaesthetics on *Haliotis midae* and their suitability for application in commercial abalone culture. *Aquaculture*, **140**, 145–51.

10 Bivalve predators and their control

Introduction

Throughout their life cycles, bivalves may contribute to the diet of a range of invertebrate or vertebrate species. Early stages, such as eggs and larvae, are especially vulnerable to predators that actively seek them as food or eat them incidentally during non-selective filtering activity. The trophic relationships between bivalves and their predators and competitors may change during the various stages of the life cycle. Also, the components of a food chain may differ according to the ecological niche in which it occurs. Thus birds, crabs, gastropods, starfish and other predators may play different roles in time and space.

Man is the ultimate predator of those bivalves that have a high intrinsic value to him, especially as food. His interest in them has created the development of cultivation practices that are aimed at optimising his investment in time, materials and resources. Past bivalve cultivation practices have often been directed towards improving natural resources by encouraging or collecting spatfalls and removing competitors and predators from the shellfish beds. These management practices have led to the development of successful fisheries for flat oysters in England and the Netherlands, cupped oysters in France, mussels in the Netherlands, France and Spain, clams in France, Spain and Italy and scallops in France and Scotland.

Over the past 30 years, the development of commercial hatcheries in Europe, which now annually produce hundreds of millions of juvenile bivalves for the shellfish industry, has required cost-effective methods to ensure the best possible survival of their valuable crops. On-growers of hatchery seed tend to use methods that exclude, rather than control, predators. Crops are kept in trays or bags, or behind fenced barriers or beneath plastic netting, often until they are of a size at which they are relatively safe from predators. Cultivators who utilise natural stocks of shellfish, such as mussels and oysters, re-laid onto ground plots, may have of necessity to control predators such as crabs, starfish and tingles by trapping or catching them.

This chapter constitutes a review that is inevitably historical and highlights the mortality inflicted by predators on molluscan shellfish of commercial importance in northern Europe, and shows how their impact has been minimised by management practices that have not always involved predator control. It deals with six major groups of animals that can be significant predators of bivalves. They are birds, fish, crabs, starfish and sea urchins, molluscs and flatworms.

Birds

Eider duck (*Somateria mollissima*) and oystercatcher (*Haematopus ostralegus*) are the most important and specialised avian predators of marine bivalves in Europe, although small numbers may also be eaten by herring gulls (*Larus argentatus*) and crows (*Corvus corone*), which open mussels, mainly by dropping them from some height onto hard surfaces.

Eider duck

Populations

Eider duck, the most common sea duck in Britain with a wintering population of 72 000 (Baillie, 1986) and in Europe, has mainly an Arctic distribution and is on the southern edge of its range in Britain. Its distribution in Britain is mainly northerly with 80% of the winter-breeding population occurring in Scotland, 17% in England and Wales, and 3% in Ireland. The European winter population is estimated at near to 2 million birds. Thus the population of the British Isles is about 4% of that in Europe.

In Scotland, eiders are common on the west coast and in Orkney and Shetland, but the largest colonies are on the east coast, especially in estuaries such as the Ythan, Eden, Tay and Montrose Basin, where they feed on natural beds of mussels. The Ythan estuary, containing about 30 ha of intertidal and shallow sublittoral beds of mussels, supports the largest mainland concentration of eiders in Britain, often with more than 4000 birds during the breeding season. More than 60% of the diet of adult eiders consists of mussels, but shore crabs, periwinkles and fish offal are also eaten. Eiders feed by diving, to depths of down to 10 m, to reach submerged mussels on intertidal or subtidal beds or on suspended ropes at mussel farms. There is a seasonality of abundance of eiders, with peaks occurring in the autumn and spring when predation of mussels is highest. It has been observed that eiders removed 4360 mussels per m^2, of 10–25 mm shell length, over a 60-day period during winter which can result in an 80% loss from November to April (Rafaelli *et al.*, 1990).

In the Wadden Sea, cockles and mussels comprise 75–80% of the eider diet. The eider population there, numbering about 250 000 birds in 1987, has altered little since the late 1970's despite increasing production of cultivated mussels. About 90–130 000 t of mussels are harvested each year from the Wadden Sea, mostly from cultivated beds. About 37% of the eiders are found near to cultivated mussel beds, which provide about 20% of the eider food, equivalent to 30 000 t (live weight) of mussels per year. There is a scientific view that eiders have little conflict with the natural mussel fishery, taking small mussels 2–3 cm in size rather than the 5 cm or larger mussels caught by the fishery, in a situation in which both the eider population and the shellfishery were still expanding.

In recent years, mussel farming on ropes suspended from longlines or rafts has developed in Scotland, especially in the sheltered sea lochs of the south-west and western coasts and Hebridean Islands. Current production is of the order of 1000 t per year and is growing. It has been observed that eiders either stripped a rope completely, or removed certain sizes or patches of mussels. Expanding cultivated mussel production may increase the local population of eider ducks. In Norway and Canada eider duck predation has become so severe that in some cases suspended mussel cultivation has become uneconomic.

In Scotland, studies showed that the number of birds present at a farm was related to:

- The history of cultivation at some sites which shows that there has been an increase in eider presence in successive years.
- Geographical position. Tagged eiders may move up to 50 km from their nesting sites. Thus Scottish sea lochs are potential targets from breeding centres of distribution. Some mussel farms in Norway, owing to their remoteness from major eider concentrations, have avoided damage.
- Season. Usually there is a spring/autumn peak of abundance of eider but with wide annual variation.
- Weather. Eiders tend to congregate in sheltered bays especially during stormy weather. Since many farms are located in sheltered areas, severe weather at exposed sites may increase the chances of eiders finding newly established farms.

Protective measures

Methods of protection depend upon the type of cultivation, i.e. longline or raft.

Physical barriers

For longlines, floating horizontal nets on the seaward side of mussel ropes and vertical surface barriers proved to be effective eider deterrents in sheltered waters. At exposed sites, however, the nets frequently became damaged and, as a consequence, were less effective.

For rafts, vertical netting, deployed to a depth equal to, or slightly deeper than, the mussel ropes, may be an effective eider deterrent, but it quickly becomes fouled. Brightly coloured net with large mesh (70 mm) reduces the likelihood of foraging birds becoming entangled. Plastic skirting, composed of flapping strips of thin ribbon-like material suspended from the raft to just below the water surface, may scare eiders by visual disturbance, flapping noise or occasional contact.

Scaring methods

A number of bird-scaring strategies have been tried with varying amounts of success. Undirected shooting and bird-scaring cartridges are effective especially when used early in the morning when the birds approach the ropes. Automatic gas guns, mounted on rafts, may be effective in the short term.

Sacrificial mussels

Mussel ropes, deployed on longlines near to rafts or on the outer fringe of the raft, may be sacrificed as food to eiders in order to preserve the bulk of the stock. The ratio of sacrificial mussels to farmed ones must be sufficiently low to be cost-effective.

Conflicts of interest may arise between farming activities and nature conservation. Consultation between the groups can lead to a compromise of working practices. A good example of this is the guidelines relating to the control of eiders on mussel farms, produced jointly by the Nature Conservancy Council for Scotland, the Royal Society for the Protection of Birds, and the Scottish Shellfish Growers Association.

These guidelines recommend that:

- The numbers of eiders should be recorded before and after the farm is established to show the times of their occurrence. Work schedules should be organised so that workers are present during periods when the eiders are likely to be there.
- Protective measures should be adopted from the outset, since eider numbers tend to increase with successive years of cultivation at a farm.
- Repetitive scaring should be arranged, since the farm will be visited by different groups of eiders.
- Shooting may be used only as a last resort.

- Rafts or grouped longlines are more cost-effective to protect than dispersed longlines.
- Larger farms are likely to be more cost-effective to protect than smaller farms.
- Barrier nets have proved to be the most effective way of excluding eiders; and
- Plastic fringes around rafts and sacrificial crops may help in the short term.

Oystercatchers

Oystercatchers are widespread in winter around the coasts of the British Isles, with a population estimated at 280 000 birds, representing about 26% of Europe's oystercatchers (Davidson *et al.*, 1991). They breed in many areas of the British Isles, especially in the north, but occur in greatest numbers in estuaries and bays during autumn and winter, outside of the breeding season. The wintering flocks contain British birds and many that breed in the Faroes, Iceland and Norway. In Europe there are few areas that support large numbers of wintering waders. The Wadden Sea, with by far the largest area of estuarine habitat in Europe (about 4000 km^2 of tidal flats), and the Dutch Delta area are of major significance with mid-winter populations of about a million waders.

The wintering populations of oystercatchers in western Europe and Britain increased in size between 1971 and 1987 owing to an expansion of breeding range and diversification of habitat use. Oystercatchers feed on various intertidal invertebrates, including ragworm and shore crabs, but the bulk of the diet in most areas consists of bivalve shellfish, mainly mussels, cockles and Baltic tellin, on which they feed when the beds are exposed at low tide. Cockles and mussels are relatively thick-shelled prey that other waders are incapable of opening. The oystercatcher, however, has a powerful beak and specialised behaviour, and is well adapted to this purpose. Oystercatchers are most numerous where their prey is most abundant in the intertidal zone. Six estuaries hold over 50% of Britain's population of oystercatchers, and these flocks can have a substantial impact on the numbers of mussels and cockles on natural beds. It has been estimated that an oystercatcher ate 270–400 mussels during a 9-hour daytime feeding period in one tidal cycle in winter on the natural beds in the Ythan Estuary, Scotland, and on average 574 mussels (mean shell length 25.7 mm) and 186 mussels (length 37.5 mm) during daylight feeding during the low water period in winter at two sites on the River Conwy. This represents the consumption of about its own body weight of wet mussel flesh per day.

On the River Exe, where up to 3500 oystercatchers may overwinter, approximately 20–30% of the million mussels 45–65 mm in size are eaten during the winter period. In the Dutch Wadden Sea, 40% of the mussels above 40 mm were cropped by oystercatchers from beds near to the Frisian island of Schiermonnikoog. This impact was considered to be representative of the Wadden Sea as a whole. Studies in the River Exe showed that the sizes of mussels selected depended on the method used to open them, and that optimal foraging was achieved when the oystercatchers changed to feeding on small mussels in spring, when the flesh weight of the larger mussels dropped owing to spawning.

In the past, there has been concern that oystercatchers could have considerable economic effect on the cockle industry in parts of England and Wales. Culling methods were adopted in the Burry Inlet, South Wales, to protect cockle stocks from oystercatchers that had increased from 8000 to 15 000 birds. Despite a management plan, which allowed 10 000 birds to be shot in 1973 and 1974, the cockle population dropped alarmingly for unknown reasons. A re-analysis of the data suggested that the impact of oystercatchers on the cockles had been overstated.

Control measures, which reduce the impact of oystercatchers feeding on cultivated mussels, largely relate to the adoption of sensible management practices, which may be a compromise between expedience and convenience. Bottom mussel culture is largely subtidal and is beyond the reach of oystercatchers. In some areas, where the lower intertidal zone is reseeded with mussels, the risk from predation by oystercatchers is increased. Oystercatcher predation of re-laid mussels in the Menai Strait was estimated to be about 10% in the winter. Cultivation in the River Exe entails the re-laying of marketable mussels from subtidal beds to the mid- to upper-intertidal zone, for convenience prior to sale. The local farmer estimated that this practice was accompanied with 30% loss of mussels from oystercatcher predation.

Fish

Flatfish such as plaice (*Pleuronectes platessa*), flounder (*Platichthys flesus*) and dab (*Limanda limanda*) may crop siphons from burrowed bivalves or eat the whole animal. In the southern North Sea, the Baltic tellin (*Macoma balthica*) is an important food for plaice, as in Scotland *Tellina* is for dab and plaice, and in England and Wales cockles are for flounder and mussels are for plaice and flounder.

Some sparid fish, such as the gilt-head sea bream (*Sparus aurata*) and white-head sea bream (*Diplodus sargus*) in Spain and France may cause damage to mussel crops, which may be sufficiently serious to justify the enclosure of the raft with nets.

In some parts of the world (e.g. Australia), shellfish farms may be enclosed with wooden or plastic fences or barriers against predation by rays. This problem is encountered to a small extent on the south-west coast of France where oyster beds may be attacked by the common sting ray (*Dasyatis pastinaca*) and the eagle ray (*Myliobatis aquila*).

Crabs

World-wide, many species of crabs are notorious predators of commercial bivalves. In Europe several species, including the swimming or velvet crab (*Necora (Liocarcinus) puber*), the brown or edible crab (*Cancer pagurus*) and especially the shore or green crab (*Carcinus maenas*) may have a considerable impact on natural and cultivated stocks of bivalves.

The shore crab is widespread and abundant in estuaries and coastal waters of Europe and parts of the Atlantic coast of North America. Shore crabs migrate up and down the intertidal shore with the flood and ebb tides. In winter most crabs remain in the sublittoral zone, but some onshore foraging may occur except in the coldest months. The number of crabs, mostly of 20–50 mm carapace width, moving shorewards on flood tides in summer in the Menai Strait, averaged 84 per m width of tideline, with densities of about two crabs per m^2 at high tide (Dare & Edwards, 1981).

Shore crabs prey heavily on young bivalves and cause significant losses to natural stocks of cockles, soft-shelled clams (*Mya arenaria*), palourde (*Tapes decussatus*) and mussels, as well as to experimental and commercially cultivated stocks of American hard shell clams (*Mercenaria mercenaria*), Pacific oysters, Manila clams, flat oysters and scallops. Shore crabs are important in controlling the abundance and distribution of various bivalves and other invertebrates such as polychaete worms, barnacles, other small crustaceans and gastropods, and they may prevent the establishment of bivalve beds by their predatory activity on the spat. Many studies have been made of the feeding behaviour and food consumption of crabs in the field and in the laboratory. Relationships for crab and prey size have been produced in Europe for shore crabs eating mussels, American hard shell clams, Pacific oysters, cockles, Manila clams, and for the brown crab eating scallops. Some of these relationships are shown for mussels (in Fig. 10.1) and for various other commercial bivalves (in Fig. 10.2).

The daily consumption rate of crabs varies widely, depending upon the size of the bivalves, the size of the crab and other biological and environmental factors. In laboratory experiments medium (45 mm) and large (75 mm)

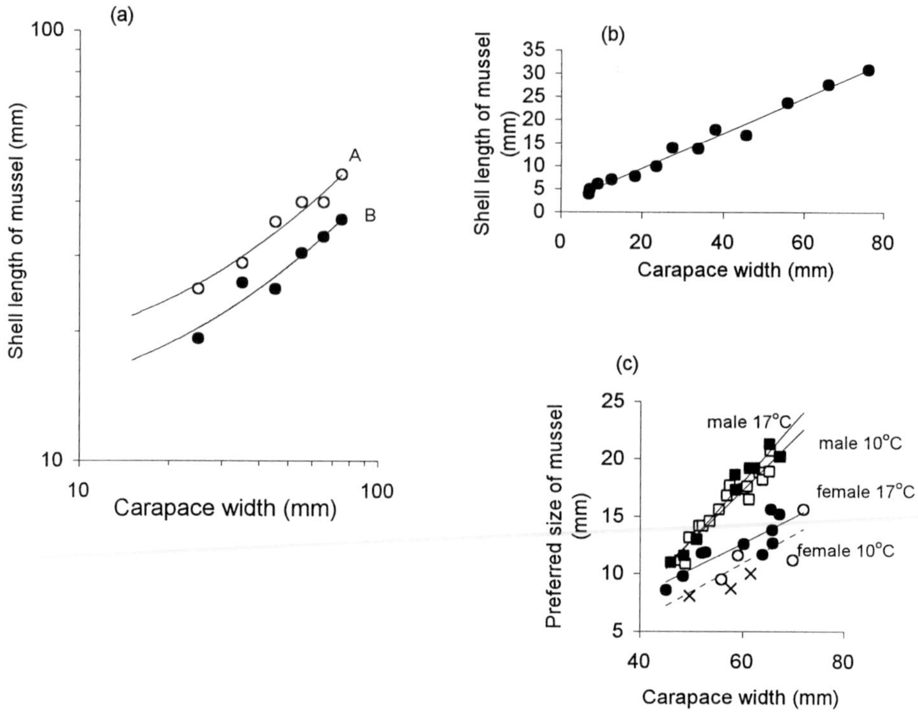

Fig. 10.1 Relationships between the sizes of *Mytilus edulis* eaten by various sizes (carapace width) of *Carcinus maenas*, redrawn from: (a) Dare *et al.* (1983), where A = the largest mussels that could be opened in laboratory experiments and B = the largest mussels consumed at a rate of two per crab per day; (b) Walne & Dean (1972), for mussels eaten at a rate of one per day; and (c) Elner (1980), for preferred sizes of mussels eaten by male (square symbols) and female (round symbols) crabs at 10 (open symbols) °C and 17° (filled symbols) C; data for sub-optimally feeding males (×) are also shown.

shore crabs ate 20–40 mussels of 20–25 mm size per day but only 2–3 Pacific oysters of this size. Another study showed that pairs of male and female shore crabs of 55–72 mm carapace width ate 75–100 Manila clams (7.5–10 mm size) per day. In the wild, geographical location of crabs or prey, level of tidal exposure and other factors can influence predation rates.

Several authors have suggested that crabs preferred sizes of prey that enable them to forage optimally on the shore. The preferred sizes of mussels (Fig. 10.1) and cockles (Fig. 10.2) are much smaller than the maximum sizes that can be eaten by crabs of given sizes. Knowledge of the maximum sizes of bivalves that can be eaten by crabs is of benefit to the cultivator for designing predator control measures. Maximum sizes taken by the largest crabs (70–80 mm) in the population (from Figs 10.1 and 10.2) are shown (Table 10.1) for some bivalve species of commercial importance. Shore crabs of this

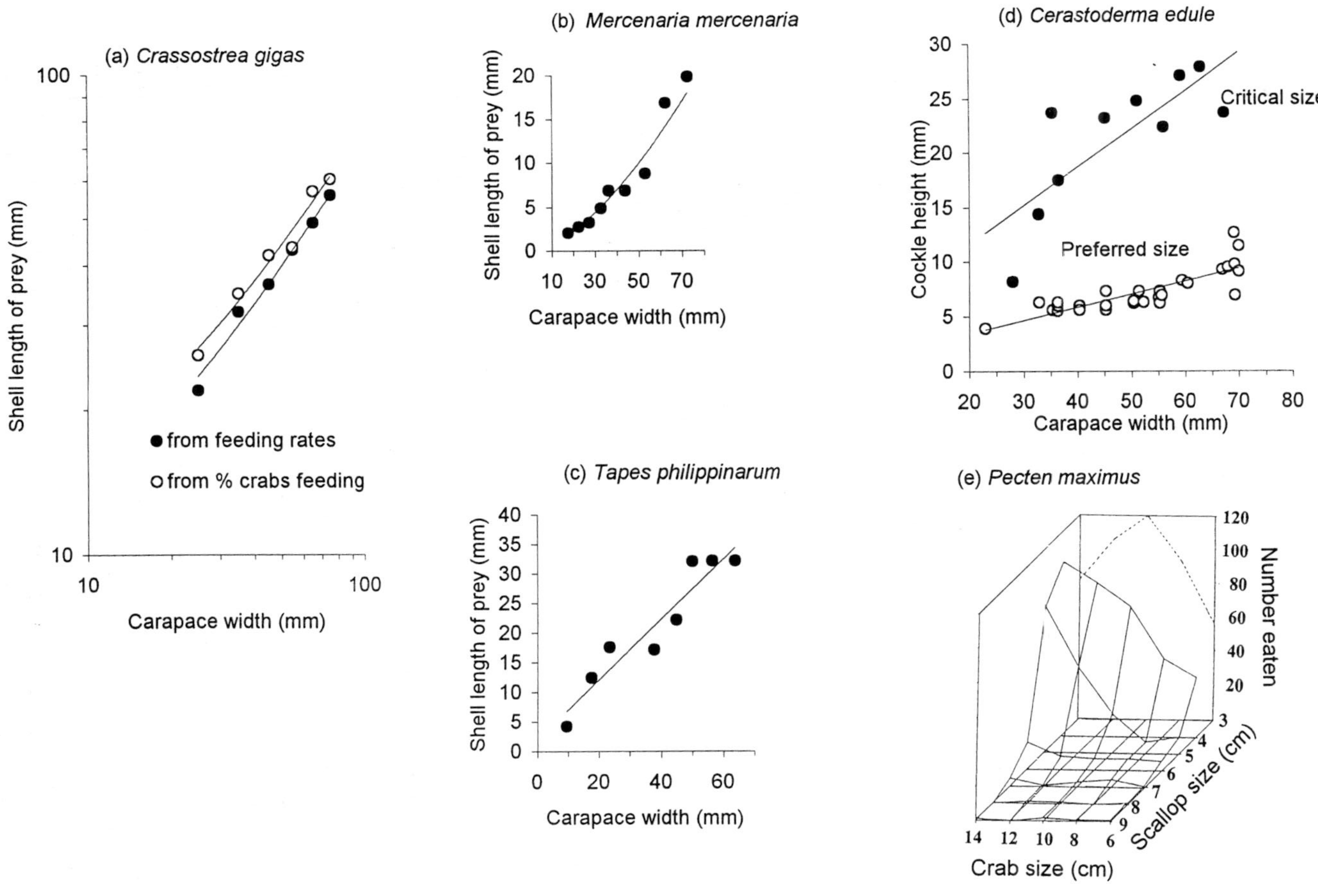

Fig. 10.2 Relationship between the sizes of crab (carapace width) and sizes of prey eaten: (a) the largest sizes of *Crassostrea gigas* that could be opened in 24 hours by *Carcinus maenas* from observations on daily consumption (●) and from % crabs feeding (○), from Dare *et al.* (1983); (b) the rate at which *C. maenas* ate at least one *Mercenaria mercenaria* per day (from Walne & Dean, 1972); (c) the largest sizes of *Tapes philippinarum* that can be eaten by *C. maenas*, drawn from Table 1 in Parache (1980); (d) the largest and preferred sizes of cockles eaten by *C. maenas* (from Sanchez-Salazar *et al.*, 1987); and (e) the numbers and sizes of *Pecten maximus* eaten by *Cancer pagurus*, from Lake *et al.* (1987).

Table 10.1 Maximum sizes of various bivalves that may be eaten by *Carcinus maenas* and *Cancer pagurus* (★)

	Maximum size eaten		
	Length (mm)	Live weight (g)	Authority
Mytilus edulis	45	10	Dare *et al.* (1983)
Crassostrea gigas	60	20	Dare *et al.* (1983)
Mercenaria mercenaria	20	3	Walne & Dean (1972)
Tapes philippinarum	37	12	Parache (1980)
Cerastoderma edule	30	9	Elner (1980)
Pecten maximus★	70 (height)	60	Lake *et al.* (1987)

size were found to be relatively scarce in a population in the Menai Strait, where most crabs (85%) were within the size range of 25–55 mm, a size considered likely to be similar to those of crabs elsewhere.

The values shown in Table 10.1, determined from laboratory experiments, provide only a guide to the vulnerability of the bivalves since, in natural conditions on the shore where alternative and easier prey may be available, crab foraging behaviour is likely to differ from that in the laboratory. To ensure high returns for effort, cultivators should seek to fully protect oysters until they reach 8–10 g, mussels until they are 35–40 mm and scallops until they reach 60–70 mm in size.

Methods of control

Shore crabs are too abundant and mobile on and around mollusc beds for them to be controlled by trapping. Choice of control method is determined largely by the value of the crop, the location of the cultivation site, the type of cultivation employed, and the aspirations of the cultivator.

Tray culture

High value crops, such as oysters, clams and scallops, may be cultivated in plastic mesh bags or pôches, trays or other containers for part or all of their cultivation cycle. The cost-effectiveness of growing them in containers is an important aspect of cultivation. The containers may be deployed on intertidal trestles or permanently suspended in the water from rafts or longlines. The intent of this type of cultivation is to exclude predators, especially crabs, from gaining access to the crop. Juvenile crabs that are sufficiently small, especially at settlement, to enter the containers can cause substantial damage to the crop if left undisturbed for prolonged periods. Regular servicing, at

Table 10.2 Survival of 10–20 mm Manila clams on intertidal plots covered with different types of plastic netting of 5 mm aperture

Barrier		Percentage survival after 2 weeks
Type	Rigidity Index*	
Tray on trestle	1250	100
Net A	500	93
Net B	30	66
Net C	12	19
No cover	0	0

* estimated as weight of plastic (in grams of plastic) per m^2 of net

monthly intervals, especially during the first year of cultivation, should minimise the impact of predation by crabs.

Net ground covers

Plastic netting is used to protect burrowing species of bivalves of hatchery origin in ground plots against predation by crabs. The net must retain the clams yet allow a good exchange of water bearing food and oxygen to ensure good growing conditions. The method has been used successfully to protect various species of clams, notably the American hard shell clam, the palourde and the Manila clam. It has been used also for protecting experimental plots of Pacific oysters, but since netting encourages siltation, which may smother non-burrowing species, its commercial application remains to be demonstrated.

Field trials in the UK, to test the growing of Manila clams to marketable size, indicate that tray cultivation is suitable only for the earliest stages of cultivation for growing the clams to about 5 mm in size, since clams larger than this develop misshapen shells. Clams larger than 10 mm grow faster, require less husbandry and develop normal shells when grown in the soil. The usual practice is to sow the clams intertidally, under netting, initially of 5 mm bar mesh size, at densities of up to 1000 clams per m^2. The net must be sufficiently rigid to prevent crabs from folding it to crush and eat the clams through the net apertures. The importance of this rigidity is shown in Table 10.2. Lighter nets, although less successful in preventing attack by crabs, can be improved, however, either by creating a gap by raising the net 5 cm above the soil or by using two layers of net. Since deployment of nets encourages siltation, both beneath and above the net, and especially so with double nets, careful management is necessary to prevent the clams from becoming suffocated. A net 2 m wide may be used to cover a plot 1.5 m in width, thus allowing the edges to be buried for good anchorage and to prevent crabs from burrowing beneath the net.

Gravel

The use of gravel or crushed stone aggregate has been suggested as a cover for American hard clam seed to provide protection from predation by crabs. The method, however, has several practical disadvantages, namely the transportation of heavy aggregate to the cultivation site and eventually the harvesting and separation of the clams from the substrate.

Gravel, 5–12 mm in size and laid at 150 t per ha, is used in France to consolidate the soil around the margins of fenced plots (Fig. 10.3), to prevent crabs from hiding in the soil and to protect the bivalves from predation by flatfish.

Protective fences

Protective fencing, used traditionally in France and the USA, can protect oysters and clams against attack by crabs. However, they can be erected only intertidally in sheltered waters without a strong tidal flow. The walls of the fence are made from 10 mm plastic netting, and are about 50 cm high with their bottoms buried to a depth of about 15 cm. The top of the fence has a crab-proof overhang made from smooth material that faces outwards at 90° or 45° to the vertical position or can be made of metal or fibreglass (Fig. 10.4).

Experimental fences, erected on muddy ground in the Menai Strait, have been shown to be effective crab barriers and, although highly successful for growing seed mussels, were less so for Pacific oysters smaller than 10 g owing to unacceptably high mortalities due to siltation. In the Menai Strait, 1 t of fenced 20–25 mm mussel seed produced about 5 t of marketable 55 mm mussels 2–2.5 years later. Unfenced mussels, however, suffer high mortalities (85%) from crab predation with 1 t of mussel seed expected to yield no more than 1 t of marketable mussels.

There are a number of practical problems associated with the use of protective fences, especially when used with mussels, which may require some husbandry management to reduce their effect. Driftweed across the fence walls and settlement of encrusting organisms may provide an easier passage for crabs to enter. Some siltation may also occur within the perimeter of the fence owing to its interference with tidal flow. The largest problem in the Menai Strait, however, was the biodeposition of 'mussel mud' with deposits of up to 17 t of mud per t of mussels grown. Its removal from within a fence between production cycles may require considerable effort.

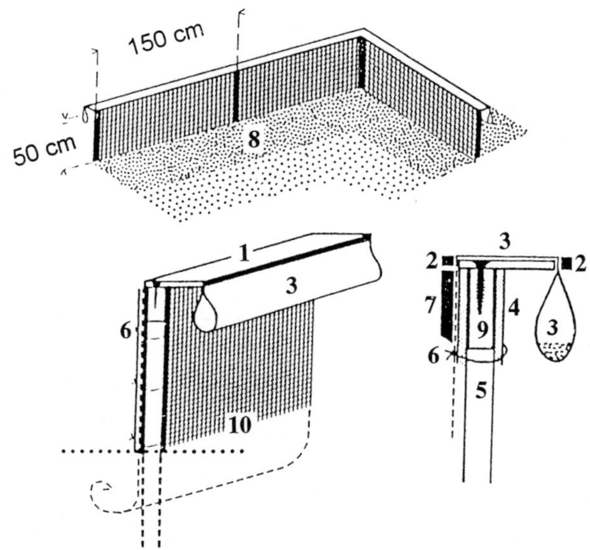

Key

1. Hardwood top board (6 cm wide, 1.5–2 cm thick)
2. Flexible plastic batten
3. Smooth plastic skirt part-filled with sediment
4. PVC tube (4 cm diameter)
5. Metal tube (3.6 cm diameter, 1.3 m high) or hardwood post (4.2 cm wide, 1.3 or 1.5 m high in muddy areas)
6. Plastic ties to secure net to post
7. Batten
8. Gravel (up to 1.5 cm in size) spread 5 cm deep for 50 cm on either side of netting
9. Wooden insert for nailing top board to metal post
10. Plastic net (up to 1 cm mesh size), 1.3 m high and with bottom edge buried in the ground

Fig. 10.3 Components of a protective fence, after IFREMER (1988).

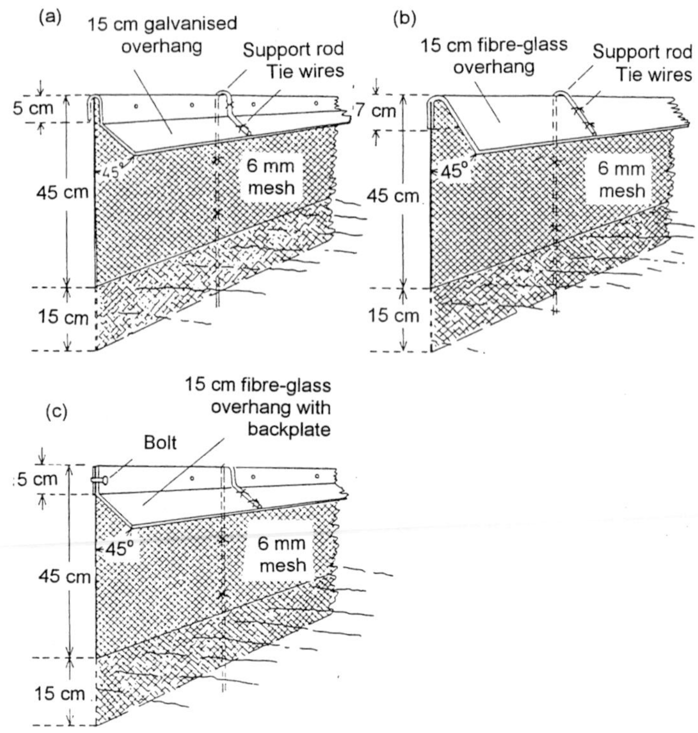

Fig. 10.4 Types of protective fences; (a) metal topped: (b) fibre-glass topped, single moulding; and (c) fibre-glass topped, double moulding (from Davies *et al.*, 1980).

Despite the relatively low value of mussels, the annual cost of fenced enclosures averaged over their life-span was found to be less than 12% and 5% of the first-sale value of uncleansed and depurated mussels, respectively (Davies *et al.*, 1980).

Starfish and sea urchins

Starfish are pests of mollusc cultivation throughout the world. In northern Europe, the common starfish (*Asterias rubens*) was considered to be the most destructive predator of natural and cultivated stocks of mussels, and on the east coast of Canada to be the worst enemy of oysters. The common starfish occurs mostly sublittorally in waters usually above 30 psu (practical salinity units), but its foraging activities may lead it into the low intertidal zone in the summer. Although its destructive power on a European oyster ground has been recorded, it caused little damage on Dutch oyster beds because of its preference for eating mussels. In laboratory trials, it ate adult flat oysters when presented as the sole food, but preferred mainly mussels and slipper limpets (*Crepidula fornicata*) when presented with a range of gastropod and bivalve species, including flat oysters. On cultivated beds in Essex, the common starfish was most likely to be found associated with large numbers of *Crepidula* (Hancock, 1955).

Dense swarms of the common starfish were recorded on intertidal beds of young mussels in Morecambe Bay, north-west England, where they reached 300–400 per m^2 and occupied up to 2.5 ha of ground. At its peak, the population contained around 2.4 million starfish which, in three months, cleared 50 ha of mussel beds estimated to have held about 3500 t of 20 mm mussels. These swarms returned to deeper water in the autumn.

Starfish have also been seen to collect in dense aggregations beneath mussel rafts in Ireland to consume those mussels that dropped off during periods of rough weather. Mussel ropes that are allowed to touch bottom may lose a large part of their crop owing to predation by starfish and crabs. Suspended mussel and scallop seed collecting bags may also collect the common starfish and the spiny starfish (*Marthasterias glacialis*) as settled post-larvae. These animals, and especially those within bags, must be removed before they grow sufficiently large to damage the crop.

The green urchin or 'burr' (*Psammechinus miliaris*) has also caused heavy losses of flat oysters in England and Scotland at sites where it has occurred in large numbers.

Methods of control

Manual removal of starfish amongst oyster spat-collecting tiles located on the low intertidal shore of Dutch oyster beds has been described as a means of controlling numbers. In the sublittoral zone, where starfish may be much more abundant and relatively inaccessible, more rapid methods of collecting them have been developed. The roller dredge (Fig. 10.5a), fished four on a beam, lifts starfish momentarily into suspension with the front roller and collects them in the rear bag. The 'starfish mop', however, incorporates a material, usually of waste cotton, in which starfish become entangled when dragged over the sea bed (Fig. 10.5b). Once on board, the starfish may be killed by immersion in hot water, fresh water, or a concentrated solution of common salt, or brought ashore to dry.

Direct application of quicklime to the beds was found to be an effective method of killing starfish in America and Canada. Although its effects on

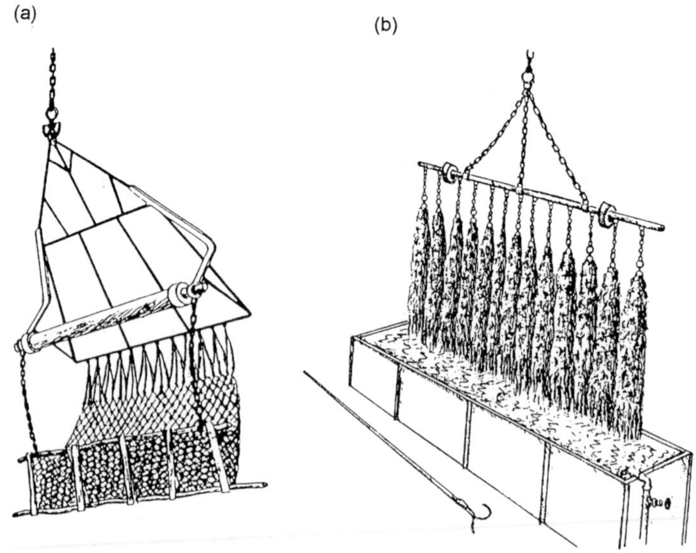

Fig. 10.5 Methods of controlling starfish: (a) starfish dredge with wooden roller; and (b) starfish mop dipped into a tank of hot water to kill entangled starfish, from Korringa (1976a,b).

oysters, and other commercial species, were considered to be negligible, its use nowadays would probably be considered to be environmentally unacceptable.

On British oyster grounds containing high concentrations of slipper limpets, fisheries advice suggested that dredged starfish should be returned to the beds as a biological control for this pest. In other circumstances, marine invertebrates, such as the red sunstar (*Solaster papposus*) and the stone crab (*Hyas areneus*), which prey on the common starfish, may also be used for natural biological control and should always be returned to the shellfish beds.

Molluscs

Important molluscan predators of cultivable molluscs are nearly all gastropods, but cephalopods (octopus) also eat mussels and oysters although their numbers are usually insufficient to cause large-scale problems. Hancock (1960) listed 27 species of gastropod predators of commercial molluscs around the world, of which five are found in British or other European waters. The European (Atlantic) species listed, with main preys, are the large necklace shell (*Natica catena*) (cockles); the dogwhelk (*Nucella lapillus*) (mussels); the European rough tingle (*Ocenebra erinacea*) (mussels, oysters); the American oyster drill (*Urosalpinx cinerea*) (oysters) and the common whelk (*Buccinum undatum*) (cockles, mussels). The common whelk uses the edge of its shell to force entry between the valves of its prey, whereas the others drill a small hole through the shell with the radula to gain access to the flesh.

The rock-shell (*Thais haemostoma*) does not occur in the British Isles but is found in the Mediterranean and Atlantic as far north as Biscay, where it feeds on patellid limpets and occasionally on mussels. On the western Atlantic coast, where it is known as the southern oyster drill, it is a serious pest of the American oyster (*Crassostrea virginica*) and the Brazilian oyster (*C. brasiliana*).

Odostomia species are small European gastropods, a few millimetres in size, which live on the shell margins of, and feed on the extruded mantle tissues of, various bivalves including scallops, queen scallops, flat oysters and mussels. Although *Odostomia* is considered to be an ectoparasite, prolonged irritation by heavy infestations can damage the adductor muscle and cause death of the host. In the UK infestation is usually at a low level, as was found with 30% of oysters at one site and 4% of mussels at another showing some degree of infestation. However, in southern Ireland it has been recorded as the most abundant gastropod on a natural bed of intertidal mussels and at a mussel farm utilising suspended ropes, where densities exceeded 200 snails per 100 cm^2 of bed.

The large necklace shell and common whelk, although common in the British Isles, are not considered to be serious predators of commercial bivalves, but the three muricid species, the dogwhelk, the European rough tingle and the introduced North American species, the American oyster drill, are more serious predators.

Most predatory gastropods detect their prey from a distance by chemoreception, with the muricid species showing acute detection of prey at a distance. The dogwhelk, the European rough tingle and the American oyster drill drill boreholes randomly on whichever prey valve is accessible.

The dogwhelk is a common intertidal species on both sides of the North Atlantic. In Europe, it occurs from the Straits of Gibralter north to the Arctic and is common around the British Isles. Since it occurs only in small numbers below low tide mark, where flat oysters are cultured, it is believed not to cause significant harm to them. Its main prey are limpets, barnacles and mussels. Dogwhelks taken from a population without access to mussels, and fed on a mixed diet of barnacles and mussels, showed preference for barnacles, but those offered only mussels developed a strong preference for this prey. In laboratory trials, adult dogwhelks each ate 2.2 mussels, and young whelks ate 1 mussel per week during the summer months (Seed, 1969). Although several hundred dogwhelks per square metre have been recorded feeding on natural intertidal beds of mussels in the spring and summer, they are not considered to be major pests of mussels. Gastropods are not considered to be important predators of mussels in the Dutch Wadden Sea or in Morecambe Bay, England. There is also no evidence that dogwhelks damage hatchery-reared oysters grown in trays in the intertidal zone. Mussel spat may settle in oyster trays and removing them can cause a considerable amount of extra work. It has been suggested that the addition of a few dogwhelks, which preferentially prey on small mussels, may help to control this problem.

The European rough tingle is found on the eastern North Atlantic coasts ranging from the Mediterranean, Madeira and the Azores to the south and west coasts of the British Isles and Ireland. It is fairly common on the south coast of Britain and rarer in the north. In a recent survey (Gibbs *et al.*, 1990), it was found to be relatively uncommon around south-west England, occurring in sheltered areas of the Channel coast and less frequently on the more exposed northern coast.

The European rough tingle is a predator of slow-moving or sessile prey, in particular barnacles, mussels and oysters. It feeds preferentially on prey with which it is familiar in its habitat, in laboratory experiments preferring mussels even when the tingles were collected from natural beds of flat oysters. Other species of commercial importance in Europe, which it ate in laboratory trials, were the Pacific oyster, the American hard shell clam, the cockle, *Venerupis* spp, the slipper limpet and other European rough tingles. It has been recorded as taking 7–13 hours to bore 0.1 mm into the shell of its prey. In laboratory trials, it was found that one European rough tingle consumed 0.2 flat oyster spat per day. Other feeding trials have indicated that the daily rate of consumption of flat oyster spat (15–20 mm in size) by the European rough tingle was nil at 10°C and 0.11 at 20°C.

The European rough tingle was abundant in all major oyster-producing areas of England prior to 1939. Severe winters since that time have severely depleted its numbers in the shallow east coast oyster grounds. However, it remained abundant on the south coast, especially in the Solent, where Europe's largest natural beds of flat oysters are to be found. In 1976, following large spatfalls of oysters in parts of the Solent, it was estimated that up to 30 million European rough tingles occurred on two oyster beds, with densities up to 90 tingles per m^2. The mortality caused by these tingles, by setting out bags of 30–45 mm oysters on the sea bed for 2 summer months, was estimated to be 11%.

The American oyster drill was introduced unintentionally into English waters from the eastern United States with deposits of the American oyster.

It was first identified on the Essex oyster grounds in 1927, with the first record dating from a collection preserved in 1920. Its presence in English waters demonstrated the risks associated with the introduction of alien species without appropriate quarantine procedures.

The American oyster drill became established in several Essex estuaries, including those of the Rivers Colne, Blackwater, Crouch and Roach, spreading from the sites of original deposits of American oysters at West Mersea and Brightlingsea by normal trade in oysters. Natural dispersion has been limited and slow. Its range has not extended northwards beyond Essex, or southwards beyond the north coast of Kent.

The voracious appetite of the American oyster drill for young flat oysters was responsible for the destruction of large quantities of thumbnail-sized spat in Essex. It has been estimated that around 50–75% of the spatfall in Essex rivers was lost to the American oyster drill, with an individual tingle capable of consuming 20 spat per season, and that more than 0.5 million oyster spat were drilled in only a few months on one ground.

The American oyster drill became especially abundant in the Rivers Crouch and Roach where densities of up to 20 000 tingles per ha were recorded in the 1950s and with similar numbers in 1972–73. Subsequent assessments of its status in the Essex rivers showed that it is in marked decline. Benthos surveys in the River Crouch in 1987–89 did not reveal any American oyster drills. Fifty tile traps set out and visited regularly at Goldhanger Creek, River Blackwater, a site where the American oyster drill once commonly occurred, caught just 142 tingles between 1985 and 1991. Of 73 individuals sexed, 31 females showed advanced imposex (the development of male sex organs in the female) resulting from the effects of absorbing tributyltin (TBT) from marine antifouling paints (Gibbs *et al.*, 1991). This phenomenon, together with a 90% decline in natural production of flat oysters in Essex since the severe winter of 1962–63, has probably been responsible for the major decline of the American oyster drill in Essex.

Methods of control

The severity of infestation of the American oyster drill on American and English grounds stimulated the development of various control measures to reduce their impact.

Chemical treatments

Research effort in the 1940s to 1970s in the USA was directed towards the development of chemical treatments, largely as disinfectants to aid movement of oysters to tingle-free areas. Over 1000 compounds were screened in controlled laboratory conditions. Dips, consisting of solutions of simple inorganic compounds such as copper sulphate, sodium chloride, and organic compounds such as formalin, were often effective against tingle spawn and other pests or fouling organisms (e.g. slipper limpets, sponges and tunicates) but not against the adult tingle.

Eventually, barriers impregnated with inhibiting chemicals were tested, which deterred tingles and starfish from crossing them. The intent was to surround a bed of oysters, etc with a strip of inert carrier such as sand or gravel mixed with about 5% of a heavy oil such as orthodichlorobenzene. Trials in the laboratory and field indicated that barriers remained effective for up to a year, although deposits of silt on top of the barrier eventually rendered them ineffective. More sophisticated formulae, containing insecticides such as lindane and Sevin, were incorporated to improve the barriers' effectiveness.

This philosophy was also applied to the general treatment of shellfish grounds, whereby mixtures of sand and chemicals were spread over the grounds to kill drills and other oyster enemies. A particularly effective treatment was Polystream, a mixture of polychlorinated benzenes, which reduced tingle catches from 16 to less than one per tile trap, with its effectiveness lasting for up to 2 years in the field trial. Chemical treatments gained some popularity with the shellfish industry in the USA, but eventually were discontinued because of concern for their effect on the marine environment. Nowadays the use of toxic chemicals in the sea is closely controlled and it is unlikely that permission would be granted to deploy them in the UK under statutory provisions of the Food and Environment Protection Act, 1985.

Catching and trapping

Mechanical or suction dredging, especially when used in the summer when the tingles are not hibernating, has been employed for their capture. A more effective method, however, utilised the summer onshore migrating behaviour of tingles for their capture. Curved roofing tiles, laid in rows at low water of spring tides, or in wired bundles in the sublittoral zone, were effective traps since they offered shelter for the tingle and a surface area on which to lay their spawn. As an example of their catching power, 225 tiles set out in an Essex river and examined on eight occasions trapped 4000 tingles and 30 000 egg capsules in one summer.

Although the European rough tingle may also be trapped by tiles, their deployment on the natural oyster fishery in the Solent was considered unlikely to be effective because of the lack of onshore migration by this species. It was also thought to be impractical because of the vast scale of the grounds (over 18 000 ha), the fast tides, the trawling activities of the fishing fleet and the large and widespread population of tingle that occurred there (Key & Davidson, 1981).

Statutory controls

Statutory powers, designed to control the introduction and spread of molluscan shellfish pests and diseases in England and Wales, were, until 1993, provided by the Molluscan Shellfish (Control of Deposit) Order, 1974. This Order had been effective in controlling the further spread of the slipper limpet, 'red worm' (*Mytilicola intestinalis*), the American oyster drill and, since 1982, *Bonamia* from areas of infestation. The Order divided the whole coastline of England and Wales into 27 designated areas within which it was illegal to deposit molluscan shellfish between designated areas, except under licence. In 1993, this legislation was superseded by the Fish Health Regulations 1992, introduced in support of the EU Directive 91/67/EEC. The new legislation disregarded the pest status of the slipper limpet, 'red worm', and the American oyster drill, and movement controls within and between EU countries were based on the presence of the bivalve pathogens, *Bonamia ostreae* and *Marteilia refringens*. *M. refringens* is not known to occur in the UK, whilst *Bonamia* occurs in three areas of England, notably in Cornwall (Dodman Point to the Lizard lighthouse), Dorset/Sussex (Anvil Point to Selsey Bill) and Essex/Suffolk (Barge Pier, Shoeburyness to Landguard Point, Felixstowe).

The decline in numbers of predatory gastropods in some areas of Europe in recent years is probably in part associated with the presence of tributyl tin in the water. In the UK, imposex has been observed in predatory drills, such as the dogwhelk, the rock shell and the American oyster drill. This chemical has now been banned as a component of antifouling paints for use on small boats in many countries in Europe and elsewhere, and the demise of the predatory drill may thus be shortlived. Recolonisation of areas where numbers of drills have been depleted, from neighbouring clean areas, may take some time because dogwhelks, rock-shells and American oyster drills lay their spawn as egg capsules in which development takes place up until the emergence of the crawling juveniles. It is unclear as yet whether the fate of the American oyster drill in England has been sealed. This drill has been severely affected by imposex in the River Blackwater and probably in the River Crouch, but it is unknown whether or how quickly unaffected populations outside or within these estuaries are able to act as centres for repopulation. The rock-shell, however, has planktonic larvae and recolonisation of affected areas from outside could be relatively quick.

Flatworms

Polyclad turbellarian flatworms may be serious predators of molluscs, including commercial species, world-wide. They may attack oysters, mussels, clams, slipper limpets, barnacles and other marine invertebrate species. In the Mediterranean Sea, the flatworm, *Stylochus mediterraneus*, was responsible for the loss of up to 90% of small oysters and occassionally preyed on young Mediterranean mussels (*Mytilus galloprovincialis*) in Yugoslavia. *S. mediterraneus* is also an important predator of Mediterranean mussels on natural beds on the Tuscan and Adriatic coasts of Italy, where it is considered as a pest by mussel farmers (Galleni *et al.*, 1980). It is not possible to control the flatworm's infestation of natural shellfish beds, but its incidence on mangrove oysters (*Crassostrea rhizophorae*), suspended from rafts was reduced by fortnightly aerial exposure, or controlled by dipping the oysters in hypersaline solutions or solutions of calcium hypochlorite (Littlewood & Marsbe, 1990).

Conclusions

In compiling this chapter, it has become clear that the role of the predator may change with time. Eider duck and oystercatcher populations in Europe have in general increased in size over the past 30 years or so. The coincidence of increased eider predation with the expansion of mussel farming in Scotland has led to the preparation of an agreed set of guidelines with which to confront the problem. Crabs and starfish appear to be ever present on some shellfish grounds, although their abundance may change owing to various environmental or biological factors. Severely cold winters in areas with shallow inshore waters and estuaries may reduce crab populations to a level that is sufficiently low to allow a resurgence of bivalve populations. American and European tingle and dogwhelk have declined in some areas of England in recent years. This may have been due partly to the presence of tributyl tin in the water and partly to the fluctuations in population numbers of oysters, mussels and other components of their natural food.

Periods of abundance of a predator have stimulated the development of methods to reduce its impact to an acceptable level. Protective fences in France are an accepted method for controlling mainly crab predation. Tile traps and dredging were used extensively to control American tingle on the fairly discrete, shallow oyster grounds in the Essex creeks. This method, however, is impractical for use on large areas of natural beds of oysters, such as those which occur in the Solent on the south coast of England. Dredging was also found to be effective in controlling starfish on Dutch and English shellfish grounds. Some chemical treatments were found to be effective in controlling starfish and tingle but their use nowadays is difficult to justify environmentally and is unlikely to gain official sanction in some countries owing to stringent legislative controls. Costly hatchery-produced bivalves and naturally caught seed of valuable species such as scallops are usually grown in plastic mesh containers, either to marketable size or until large enough to withstand predation on release to prepared areas of the sea bed. Hatchery-reared clams, however, require special protection with plastic netting of sufficient rigidity to prevent crabs from manipulating and breaking them open through the net.

Many European countries, including those belonging to the EU, now have statutory powers to control the introduction of 'non-native' marine organisms, including competitors, pests and predators, for deposit in their waters. Those countries who may wish to introduce a species from overseas are advised to work to a set of guidelines produced by the International Council for the Exploration of the Sea (ICES), including quarantine procedures, to reduce its potential impact in its new environment (ICES, 1995). The UK has statutory powers to control the deposit of molluscan shellfish from overseas and between areas of its own coastline. This legislation has, in the past, helped to control the spread of pests and diseases such as the slipper limpet, 'red worm' and American oyster drill and *Bonamia*. The new legislation, introduced in 1993 [based on EEC Directive (NO. 91/67/EEC)], has aimed at simplifying and standardising the control of movements of fish and shellfish between member countries by allowing deposits from zones of similar or lesser pest or disease status. It has, however, been weakened by its disregard of the slipper limpet, 'red worm' and American oyster drill, as pests. Its performance in the future will be judged on the impact this omission has on the shellfish industry.

Although levels of predation may fluctuate over the years, the cultivator must remain vigilant to ensure the safe survival of the crop by adopting control methods or management strategies that reduce its impact to an acceptable level.

References and bibliography

Baillie, S. R. (1986) Eider. In: *Atlas of Wintering Birds in Britain and Ireland* (comp. P. Lack), 114, T. & A. D. Poyser Ltd, Carlton, Staffordshire.

Dare, P. J. (1976) *Settlement, growth and production of the mussel,* Mytilus edulis *L. in Morecambe Bay, England.* Fishery Investigations, MAFF, London, Series 2, **28** (1), 1–25.

Dare, P. J. & Edwards, D. B. (1981) Underwater television observations on the intertidal movements of shore crabs, *Carcinus maenas*, across a mudflat. *Journal of the Marine Biological Association of the United Kingdom*, **61**, 107–16.

Dare, P. J., Davies, G. & Edwards, D. B. (1983) *Predation on Juvenile Pacific oysters* (Crassostrea gigas *Thunberg) and mussels* (Mytilus edulis *L.*) *by Shore Crabs*, (Carcinus maenas *(L.)*). Fisheries Research Technical Report, MAFF, Directorate of Fisheries Research, Lowestoft, **73**.

Davidson, N. C., Dd'A. Laffoley, J. P. Doody *et al.* (1991) *Nature Conservation and Estuaries in Great Britian.* Nature Conservancy Council, Peterborough.

Davies, G., Dare, P. J. & Edwards, D. B. (1980) *Fenced Enclosures for the Protection of Seed mussels* (Mytilus edulis *L.*) *from predation by shore crabs* (Carcinus maenas *(L.)*). Fisheries Research Technical Report, MAFF, Directorate of Fisheries Research, Lowestoft, **56**.

Elner, R. W. (1980) The influence of temperature, sex and chela size in the foraging strategy of the shore crab, *Carcinus maenas* (L.). *Marine Behaviour and Physiology*, **7**, 15–24.

Galbraith, C. (1992) *Mussel Farms: Their Management alongside Eider Ducks*, 2nd edition. HMSO for Scottish Natural Heritage, Edinburgh.

Galleni, L., Tongiorgi, P., Ferrero, E. & Salghetti, U. (1980) *Stylochus mediterraneus* (Turbellaria: Polyclada), predator on the mussel *Mytilus galloprovincialis*. *Marine Biology*, **55**, 317–26.

Gibbs, P. E., Spencer, B. E. & Pascoe, P. L. (1991) The American oyster drill, *Urosalpinx cinerea* (Gastropoda): evidence of decline in an imposex-affected population (R. Blackwater, Essex). *Journal of the Marine Biological Association of the United Kingdom*, **71**, 827–38.

Great Britain in Parliament (1985) *Food and Environment Protection Act* 1985. Chapter 48. HMSO, London.

Hancock, D. A. (1955) The feeding behaviour of starfish on Essex oyster beds. *Journal of the Marine Biological Association of the United Kingdom*, **34**, 313–31.

Hancock, D. A. (1960) The ecology of the molluscan enemies of the edible mollusc. *Proceedings of the Malacological Society of London*, **34** (3), 123–43.

Hancock, D. A. (1974) *Oyster Pests and their Control.* Laboratory Leaflet, MAFF Directorate of Fisheries Research, Lowestoft, **19**.

ICES (1995) *Code of Practice on the Introductions and Transfers of Marine Organisms.* ICES, Copenhagen. Cooperative Research Report No. 204.

IFREMER (1988) *La Palourde – Dossier d'Elevage.* IFREMER Documents and Publications Unit, Plouzane, France.

Key, D. & Davidson, P. E. (1981) *A Review of Development of the Solent Oyster Fishery, 1972–80.* Laboratory Leaflet, MAFF Directorate of Fisheries Research, Lowestoft, **52**.

Korringa, P. (1976a) *Farming Marine Organisms Low in the Food Chain.* Vol.1. *A Multidisciplinary Approach to Edible Seaweed, Mussel and Clam Production.* Elsevier, Amsterdam.

Korringa, P. (1976b) *Farming the Cupped Oysters of the Genus Crassostrea.* Vol.2. *A Multidisciplinary Treatise.* Elsevier, Amsterdam.

Lake, N. C. H., Jones, M. B. & Paul, J. D. (1987) Crab predation on scallop (*Pecten maximus*) and its implication for scallop cultivation. *Journal of the Marine Biological Association of the United Kingdom*, **67**, 55–64.

Littlewood, D. T. J. & Marsbe, L. A. (1990) Predation on cultivated oysters, *Crassostrea rhizophorae* (Guilding), by the polyclad turbellarian flatworm, *Stylochus* (*Stylochus*) *frontalis* Verrill. *Aquaculture*, **88**, 145–50.

Orton, J. H. (1929) Habitats and feeding habits of *Ocinebra erinacea*. *Nature,* London (Ser. 2) **124**, 370–1.

Orton, J. H. (1930) On the oyster drills in the Essex estuaries. *Essex Naturalist*, **22**, 298–306.

Orton, J. H. & Winckworth, R. (1928) The occurrence of the American oyster pest *Urosalpinx cinerea* (Say) on English oyster beds. *Nature,* London, **122**, 241.

Parache, A. (1980) *Les relations 'proie-predateur' entre le crabe verte* Carcinus maenas *et la palourde* Ruditapes philippinarum. Bulletin Mensuel, Office National de la Chasse, France, 299–309.

Raffaelli, D., Faley, V. & Galbraith, C. (1990) Eider predation and the dynamics of mussel bed communities. In: *Trophic Relationships in the Marine Environment* (eds M. Barnes & R. N.

Gibson), pp. 157–169. *Proceedings of the 24th Marine Biological Symposium*. Aberdeen University Press, Aberdeen, Scotland.

Sanchez-Salazar, M. E., Griffiths, C. L. & Seed, R. (1987) The effect of size and temperature on the predation of cockles, *Cerastoderma edule* (L.) by the shore crab, *Carcinus maenas* (L.). *Journal of Experimental Marine Biology and Ecology*, **111**, 181–93.

Seed, R. (1969) The ecology of *Mytilus edulis* L. (Lamellibranchiata) on exposed rocky shores. 2. Growth and mortality. *Oecologia*, **3**, 317–50.

Spencer, B. E. (1991) Predators and methods of control in molluscan shellfish cultivation in north European waters. In: *Aquaculture and the Envirionment*. (eds N. De Pauw & J. Joyce), pp. 309–37. European Aquaculture Society Special Publication No. 16, Gent, Belgium.

Thiesen, B. F. (1968) Growth and mortality of culture mussels in the Wadden Sea. *Meddelelser fra Danmarks Fiskeri–og Havundersøgelser*, (N. S.), **6**, 47–78.

Walne, P. R. & Dean, G. J. (1972) Experiments on predation by the shore crab, *Carcinus maenas* L., on *Mytilus* and *Mercenaria*. *Journal du Conseil international pour l'Exploration de la Mer*, **34** (2), 190–9.

11 Criteria for selecting a site for bivalve cultivation

Introduction

Selecting a site suitable for bivalve cultivation is a multi-factor problem that requires an assessment of a range of information to ensure that a new business venture stands a good chance of success. To ensure that a site's economic and environmental suitability is well judged some preliminary planning, incorporating an assessment of the physical, biological, legal and economic aspects of the site, would represent time well spent. Some of the factors relevant to the United Kingdom are outlined below.

Environmental factors

Location

The suitability of a range of habitats (sea lochs, estuaries, inshore coastal sites) has been evaluated in different years with trial plantings of oysters in trays. Sites where commercial and experimental Pacific oyster cultivation have been undertaken (Fig. 11.1) are widely distributed around the coast of England and Wales. Differences in growth between sites (Table 11.1) usually reflect differences in environment that may be fairly specific to the site e.g. silt load, salinity or sea temperature, but which may vary widely between and within years.

The substrate of a site is an important consideration for bivalve cultivation. Subtidal cultivation of oysters and mussels requires a sediment sufficiently firm to prevent the stock from sinking and becoming smothered. Muddy soils may well accumulate seasonally or over a longer period of time and can be brought back into productivity by harrowing the ground before re-laying stock. Intertidal cultivation of oysters in containers on trestles requires a firm sediment to prevent the equipment from sinking, although horizontal plates fitted to the feet of the trestles reduce this effect. Clam cultivation requires a softer sediment such as mud, sands and mixtures of the two, since this is their natural habitat. Hard sediments are unsuitable, unless

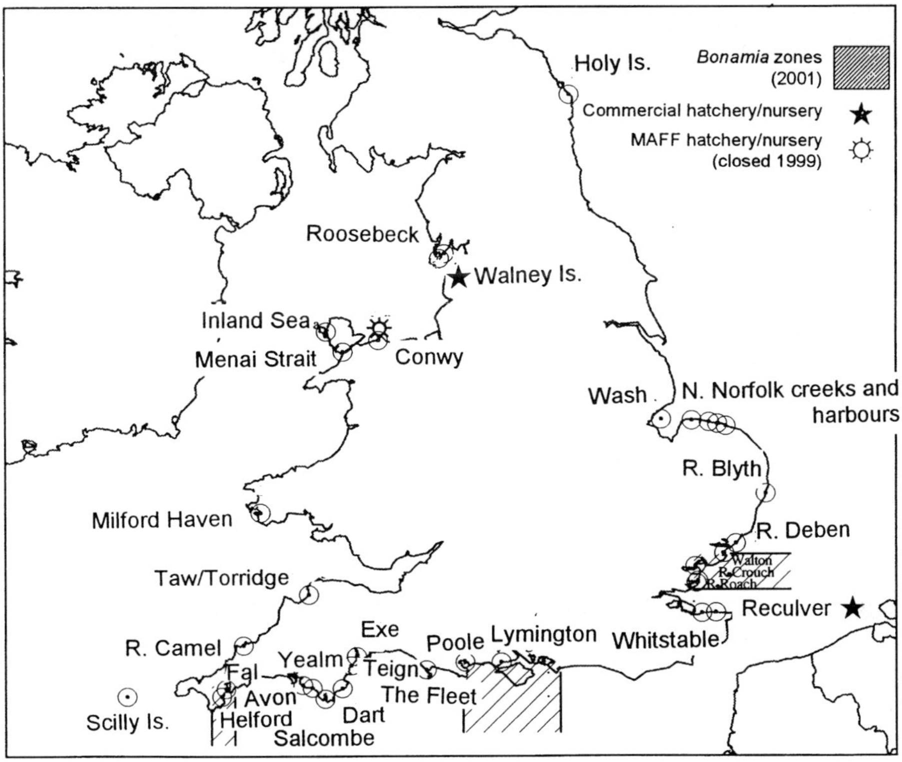

Fig. 11.1 Pacific oyster cultivation sites in England and Wales. Hatched zones show areas in which *Bonamia* infestation of flat oysters has been detected.

loosened by raking or forking before planting hatchery seed, since the clams are unable to bury themselves sufficiently quickly to avoid predation by crabs.

Sea temperature

This has a major effect on the seasonal growth of bivalves and is responsible for some of the differences between sites. Growth usually begins in April, reaching a peak in July and August, and declines to a low level by November and December (Fig. 11.2). The seasonal pattern and range of sea temperature is influenced by latitude, water depth, shelter, tidal exchange etc., and it also varies between years and between sites (Table 11.2). Shallow creeks and estuaries usually have a higher average summer temperature but a wider daily and seasonal range, than deeper, more stable bodies of water.

Table 11.1 Growth and survival of Pacific oysters in trays at various sites in the UK. The trials lasted from April to September (1975) or from April to October (1967, 1978)

Location	Maximum mean monthly temperature (°C)	Mean weight (g)	% survival
1967		15★★	
Highland, Loch Tournaig	13–14	46	95
Strathclyde, Linne Mhuirich	15–16	45	95
Anglesey (Tal-y-foel)	16–17	66	99
Anglesey (Holyhead)	16–17	77	99
Devon, River Yealm	18–19	75	95
Hampshire, Newtown River	18–19	104	100
Essex, River Crouch (raft)#	18–19	80	100
Essex, River Crouch (Althorne Creek)#	18–19	77	94
Essex, River Roach#	18–19	77	93
1975		0.6★★	
Anglesey (Tal-y-foel)	16–17	22.1	95
Devon, Kingsbridge#	15–16	11.2	88
Hampshire, Emsworth#	17–18	11.0	55
Essex, River Roach#	18–19	14.4	94
Essex, West Mersea#	18–19	14.5	100
Norfolk, Brancaster	16–17	32.4	88
1978		0.02★★	
Anglesey (Tal-y-foel)	16–17	5.4	93
Anglesey (Inland Sea)	18–20	4.6	99
Devon, Teignmouth	15–16	7.9	94
Hampshire, Hilsea Lido★	20–22	30.7	98
Essex, Goldhanger Creek#	18–19	3.1	97

★ stock planted in May, initial weight 0.004 g
★★ initial mean weight (g)
TBT affected sites

The shallow and enclosed waters of the Inland Sea (Anglesey) warm up rapidly in the spring and early summer to provide excellent nursery growing facilities (Fig. 11.2). At most sites in England and Wales, Pacific oysters should reach market size (75 g) in 4 years or less. However, in cooler waters, oysters may take an extra year or more to attain this size.

Freezing winter temperatures can kill bivalves if they are exposed in air for prolonged periods. Submerged bivalves have a better chance of survival because sea temperatures are higher than air temperatures in severely cold conditions.

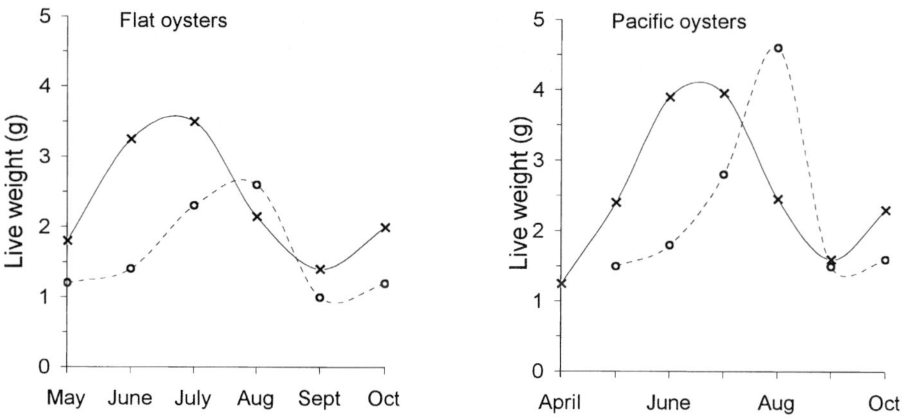

Fig. 11.2 Size reached by 1 g flat oysters and Pacific oysters in the Inland Sea (x) and Menai Strait (o), Gwynedd, in the summer of 1976.

Table 11.2 Mean monthly sea water temperatures (°C) at selected sites in the UK

	Jan	Feb	Mar	Apr	May	Jun	Jul	Aug	Sep	Oct	Nov	Dec
Loch Spelve (Mull)	7.7	7.0	6.6	7.3	8.4	9.9	11.2	12.2	12.2	11.8	10.6	9.2
Loch Sween	6.2	6.3	6.6	7.6	9.5	11.3	12.5	14.2	13.6	12.1	9.3	8.0
Loch Mhuirich	6.7			7.4	10.5	13.0	16.0	16.1	15.6	13.2	9.8	7.8
Brancaster	4.5	4.2	5.8	8.4	11.6	15.2	17.2	17.9	15.4	11.9	8.3	5.7
River Crouch			4.9	8.2	12.5	16.3	18.8	18.5	16.2	12.5	7.5	4.8
Poole	5.6	5.2	6.3	8.5	12.4	15.8	18.1	18.0	15.8	12.6	9.0	6.5
River Exe	8.8	9.1	8.6	11.0	12.0	15.0	17.0	18.0	16.0	14.0	13.0	11.0
Teignmouth	8.0	7.8	8.0	8.8	11.2	13.7	16.2	16.7	16.2	13.7	11.3	9.6
Milford Haven				8.1	11.0	14.1	15.4	16.2	16.0	14.5	13.0	
Menai Strait	6.7	6.0	6.6	8.2	10.9	13.7	15.8	16.6	15.2	12.6	9.9	7.6
Inland Sea (Anglesey)	4.8	4.5	7.3	9.5	14.6	15.8	15.9	16.2	14.1	12.7	10.5	

Water movement

Extreme wave action and tidal currents can cause physical damage to oyster-growing installations or to the oysters themselves. Sheltered areas with tidal flows of up to 1–2 knots (50–100 cm/s) provide the best conditions. However, tray cultivation can be successful in water of minimal flow (e.g. Inland Sea, where gentle wave action and rise and fall of tide are the only means of providing adequate exchange of water over the oysters). At more exposed sites (e.g. Morecambe Bay and Whitstable), robust installations are necessary to withstand the extra buffeting by the waves.

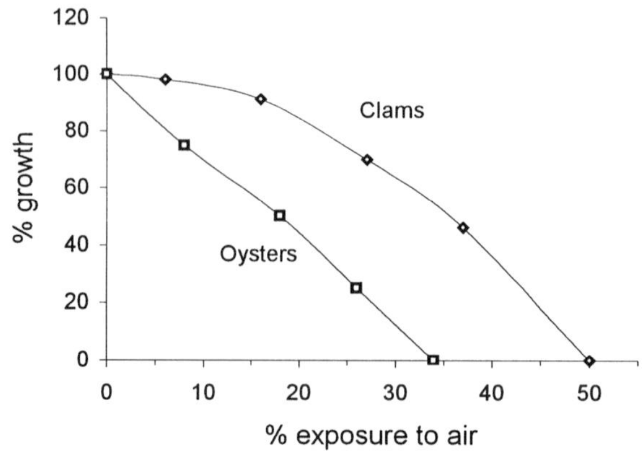

Fig. 11.3 Growth of Pacific oysters in trays and Manila clams in ground plots at various levels of intertidal exposure.

Exposure to air

Bivalve growth is strongly influenced by the length of time during which they are submerged by the tide and, therefore, feeding. Young bivalves cultured on the foreshore should be kept as near as possible to low water of spring tides to ensure nearly 100% immersion. Growth stops when oysters in trays or clams in ground plots are exposed to air for more than about 35% and 50% of the time respectively (Fig. 11.3). This fact, however, can be used to advantage by the cultivator who may wish for commercial reasons to slow down or temporarily stop the growth of his stock by moving it higher up the beach. He can do this with confidence during the summer without fear of causing undue mortalities or reducing the meat quality of his stock. This practice is routinely adopted in Korea and Japan for 'hardening off' wild-caught spat prior to sale.

The difference in sensitivity in growth to tidal exposure by the two species probably relates to the amount of desiccation caused by the method of cultivation.

Siltation

Trays, particularly those with small mesh, may quickly clog with silt. This can smother the oysters or clams or cause them to grow more slowly because of the poor exchange of water. Areas vary considerably in the silt content of their waters, but those with high silt loads can be used successfully for cultivation provided trays are serviced regularly. In very turbid waters (e.g. River Roach and West Mersea), trays may need cleaning every 2 weeks.

The placement of trestles and other structures on the foreshore interferes with current flow by reducing it significantly. At some sites, the reduced flow may cause suspended sediment to fall out of suspension and to accumulate beneath, between and around the trestles so quickly as to require regular removal to prevent the equipment and oysters or clams in containers from becoming buried. This situation is expensive to remedy and should be avoided if necessary.

Salinity

Open coastal areas are usually fully saline with small seasonal variations of between 30 and 35 practical salinity units (psu). Estuaries have a daily input of fresh water, and consequently have a continually varying salinity pattern with values ranging from 0 (fresh water) to about 35 psu, depending on their proximity to the sea, state of the tide, and rainfall. Although most bivalves grow well at a range of salinities, with mussels showing good growth between 20 and 35 psu, flat oysters, clams and scallops prefer higher levels near to 30 psu and Pacific oysters prefer lower levels near to 25 psu.

Oxygen

The oxygen content of sea water is usually sufficient for bivalves to respire normally. In some circumstances, however, (e.g. in hot, calm weather during periods of low water movement at low water of neap tides, or during periods of alga bloom decay) oxygen in sea water may become depleted. This can stress bivalves, causing them to gape and possibly to die. These periods are usually of short duration and the danger passes before remedial action, such as reducing stock density or moving stock to suitable waters, can be taken.

Pollution

Waters contaminated with heavy industrial pollution (e.g. heavy metals and organic compounds) are unsuitable for bivalve culture. Similarly, areas receiving untreated sewage should be avoided because of their high organic load and potential pathogenic virus content. It is now a statutory requirement [Food and Safety (Fishery Products and Live Shellfish) (Hygiene) Regulations, 1998] that shellfish beds must be classified according to the coliform bacterial levels of the bivalve flesh before being sold for human consumption (see Table 1.5, Chapter 1). This is a time-consuming process, undertaken by the local environment health department, which may take up to 2 years before sufficient seasonal information is gained to enable a classification to be made.

Shellfish harvested from A grade waters may be marketed live without further treatment. Those from B grade waters require depuration, while those from C grade waters must be re-laid for several months in waters of A or B classification to allow time for them to reach an acceptable bacterial standard before further treatment or marketing. Re-laying shellfish in cleaner waters, even if these are conveniently available close by, is an additional expense that can be avoided with careful site selection. Faecal bacterial content, however, is not a problem with young bivalves during the early years of cultivation, but once the stock reach edible size they must be purified in cleansing tanks before sale for consumption.

Some areas naturally contain high levels of heavy metals, (e.g. copper and zinc), washed out from disused mine workings and geological strata. These metals accumulate in the flesh without harming the bivalves but may render it unpalatable owing to its unpleasant flavour. Copper tainting, for example, has been reported from some areas of the Fal, Tamar and Lynher, but this type of contamination and others, such as oil taint, can be removed by re-laying stock in unpolluted waters for several months.

Exceptionally adverse conditions

An initial well-prepared site survey should eliminate those sites that are likely to suffer from extremes of any of the factors described above. Enquiries at Fisheries Laboratories, Sea Fisheries Inspectorate Offices, Sea Fisheries Committees (England and Wales), Public Health Laboratories, Local Authorities and Harbour Boards will often reveal information on water temperature, salinity, pollution, previous history of bivalve cultivation, etc. Nevertheless, it is wise to supplement such knowledge with trial plantings of bivalves before starting a major enterprise. The exceptional losses which occur very infrequently (for example due to flooding, storm damage, severe winters, major oil pollution, etc.) are difficult to anticipate, but insurance against such losses is possible now that aquaculture is recognised as an insurable risk.

Biological factors

Availability of seed

A bivalve cultivation business is dependent on a regular annual supply of seed. Natural stocks of seed mussels may be available locally, but spatfalls may be insufficiently regular to sustain restocking on an annual basis. In such cases it may be necessary to collect or purchase seed from distant areas that may involve expensive transport costs. In recent years Morecambe Bay, Dornoch Firth, Solway Firth and Portland Bill have provided sizeable quantities of seed mussels to the industry. The natural flat oyster fisheries of the River Fal and the Solent have traditionally supplied seed to other areas. The Essex fisheries have not been self-sustaining since the cold winter of 1962–63 killed most of the stock. However, a number of oyster fisheries located in the creeks of West and East Mersea and at Walton-on-the-Naze and in Poole Harbour continue to flourish from the deposit of half-grown Solent oysters on the beds for one growing season. A short re-laying period is necessary to reduce the risk of high mortalities of the oysters from *Bonamia* disease.

In Scotland, a scallop and queen scallop cultivation industry, based on the collection of wild seed of *Pecten maximus* and *Chlamis* (*Aequipecten*) *opercularis* is emerging, but dependence on a regular annual supply is causing the industry to turn to hatchery-reared stock to overcome this problem. Although hatchery technology is now sufficiently well advanced to provide industry with large quantities of some of the more valuable species of bivalves, such as flat oysters, Pacific oysters, Atlantic palourdes (*Tapes decussatus*), Manila clams (*Tapes philippinarum*) and American hard shell clams (*Mercenaria mercenaria*), it is not yet well developed for scallops. Pacific oysters, however, form the backbone of the cultivation industry based on hatchery-reared stock. Their small size (5–10 mm) at stocking, necessitates careful husbandry practices in the early stages of cultivation to ensure good growth and survival. The industry is largely based on the use of plastic mesh containers on trestles to hold the stock.

Food

Bivalves feed by filtering microscopic algae (phytoplankton) and organic detritus from sea water. An adult oyster may pump up to 10 l (2.2 gallons) of sea water per hour through its body cavity, depending on its size, sea temperature and other environmental and biological factors. The value of naturally occurring food is impossible to evaluate by inspection of water samples, since seasonal abundance and quality of microalgae fluctuate widely. Normally, however, peak abundance of phytoplankton occurs in the spring with a lower peak in the autumn. Coastal and estuarine waters generally have sufficient food in relation to prevailing temperature to support reasonable growth. Blooms of algae may also occur, usually in response to a combination of certain environmental conditions. These algae are often nutritionally unsuitable and consequently may reduce the growth of bivalves.

Toxic algae

Some microscopic algae produce toxins that accumulate in the flesh of mussels, oysters, clams and scallops and in some species of crustacia (swimming crabs, brown crabs, and *Nephrops*) that feed on them. Dinoflagellates belonging to the genera *Alexandrium*, *Gymnodinium*, *Pyrodinium*, *Dinophysis*, *Prorocentrum* and *Pseudonitzschia* are frequently implicated. Shellfish containing the toxins can induce paralytic shellfish poisoning (PSP, from *Alexandrium*, *Gymnodinium* and *Pyrodinium*), diarrhoetic shellfish poisoning (DSP, from *Dinophysis* and *Prorocentrum*) and Amnesic Shellfish Poisoning (ASP, from *Pseudonitzschia*) in humans who eat them. A relatively new biotoxin, azaspiracid poisoning (AZP), causing gastro-enteritis symptom's has appeared in Ireland, Scotland and Norway. Neurotoxic Shellfish Poisoning (NSP) occurs in some parts of the world but is unknown in UK waters. PSP and ASP are more dangerous, and they have caused human deaths. The toxins are not denatured by cooking or eliminated by cleansing the shellfish in depuration tanks.

The incidence of occurrence of PSP, DSP and ASP is low in England and Wales. However, in Scotland the presence of PSP and ASP was widespread in 1998–2000, leading to the closure of fisheries, especially scallops, with large economic loss (see Chapter 1). There is now a statutory requirement for sea water and bivalves from coastal areas of the whole of England, Wales and Scotland to be tested for these dinoflagellates and toxins as an early warning system. If the level of toxins exceeds a safe threshold the collection of shellfish for consumption is prohibited until the levels become safe.

Predators

Unprotected bivalves are eaten by various predators, especially shore crabs, drills or tingles, starfish and, to a lesser extent, by birds.

The shore crab (*Carcinus maenas*) is the most abundant and widespread predator in estuaries and coastal sites and has a voracious appetite for young bivalves, which it eats after cracking open the shells. The largest crabs (over 65 mm wide) can eat oysters up to about 10 g, but, since crabs of this size are relatively scarce, oysters larger than 10 g may be considered to be crab-proof. The abundance of shore crabs on the foreshore may not be obvious because of their habit of advancing up the shore on the flood tide and retreating into deeper water or hiding on the ebb.

Oyster drills or tingles are marine snails that eat bivalves by rasping a hole through the shell to gain access to the flesh. The distribution of these pests is fairly limited. The American whelk tingle (*Urosalpinx cinerea*) occurs around the Thames estuary in Essex and Kent, and the European rough tingle (*Ocenebra erinacea*) in some of the important oyster grounds of the Fal, Helford River, Solent and estuaries of the east and south coasts. It was reported that 58% of an oyster spatfall in the River Crouch was devoured by the American whelk tingle (Hancock, 1954), and in 1978 *Ocenebra* killed 11% of 30–45 mm flat oysters laid on the sea bed in a plastic mesh container in Southampton Water for 2 months in the summer (Key & Davidson, 1981). However, in recent years tingle abundance in some areas has been reduced by the presence of tributyl tin (TBT) in the water. Following the ban on the use of TBT its concentration in the environment has continued to diminish and tingles are likely to re-emerge again as important predators in the future.

The common starfish or 'five fingers' (*Asterias rubens*) can attack and eat bivalves by pulling apart the shells slightly and digesting the flesh *in situ*. It is considered to be a minor pest of oysters, however, since starfish are restricted to high salinity areas, usually in the sublittoral zone, and prefer to eat mussels and slipper limpets, which themselves are competitors of oysters. They can occur in high abundance on mussel beds and consume large numbers of stock in a relatively short period of time.

Disease

Bivalve diseases are uncommon in the UK and historically have been associated with imported stock. The only disease prevalent at the present time is *Bonamia*, that infects the blood cells of the flat oyster causing high mortalities under certain conditions of cultivation. It was introduced with unauthorised deposits of oysters in 1982 into the River Fal, where it became established in the local stocks of oysters. Before oyster movement controls could be enforced, the disease was transferred to the River Blackwater and Walton Backwaters in Essex and to Poole Harbour, River Beaulieu,

Emsworth Harbour and some parts of the Solent on the south coast. These three general areas remain the sole areas of infection, and movement controls on flat oysters are in force to prevent the further spread of the disease into uninfected areas.

Competitors

Established oyster grounds have suffered from the effect of slipper limpets (*Crepidula fornicata*), which compete with the oysters for food and space and, in silty waters, also produce a muddy substrate that is unsuitable for oyster cultivation without costly husbandry procedures. The slipper limpet was introduced accidentally into this country with imported oysters from America in about 1880 and has now spread around our coast to most oyster producing areas. It was estimated that 1150 t of slipper limpets inhabited a 5-mile stretch of the River Crouch, Essex, or about ten times the weight of oysters in that area (Walne, 1956). Since mechanical dredging and disposal is the main means of control, the costs of reclamation of grounds are likely to be high. However, in the Solent the presence of slipper limpets benefits the fishery by providing an abundant supply of cultch for the settlement of flat oyster spat.

Fouling organisms

Fouling is a general term used in aquaculture to describe animals and plants that attach themselves to immersed equipment and to the bivalves. The type and degree of fouling varies with locality and its main effects are to reduce the flow of water and food to the bivalve under cultivation in trays or on ropes and to increase the weight and drag on floating installations. Green filamentous seaweed (*Enteromorpha*) can be controlled by spraying with dilute copper sulphate solution (10% w/v), followed by exposure to air for 1 hour. However, such antifouling compounds are toxic and it is better to adopt the less harmful practice of pressure hosing, or to turn the bags over frequently. In the summer weed-fouled netting covering clam ground lays usually requires monthly removal with a scraping device such as a squeegee, but larger areas may need the use of a tractor-driven brush several metres wide (see Chapter 5).

Individual or colonial sea squirts (tunicates) can be killed, without harming the bivalves, by immersion in saturated brine solution for 5–10 minutes followed by exposure to air for one hour. Barnacle and worm encrustations are not usually a problem on bivalves but can be removed from empty containers using high pressure (2000 pounds per square inch) water sprays, or by leaving the containers ashore for several months. Barnacles can be crushed using a garden roller if the bag is sufficiently flexible, while immersing plastic trays in hot water is effective but costly. Mussel spat may also settle in bivalve trays and removing them can cause the cultivator extra work. Large settlements of mussels create competition for food and space, whereas sparse settlements cause localised clumping of bivalves that become bound together by mussel byssal threads. The addition of a few dogwhelks (*Nucella lapillus*), which preferentially prey on small mussels, may help to control this problem (Minchin & Duggan, 1989).

Legal aspects of cultivation

Public rights

The public has a common law right to fish tidal waters, in rivers and estuaries and in the sea within territorial waters (12 nautical miles from baselines), except where a private property right has been obtained which excludes the common right. Subject to certain constraints, anyone can grow molluscs where a public right of fishery exists. The constraints are:

- The public right of fishery must not be interfered with.
- If a Regulating Order for a specified species of bivalve exists, then they may only be taken in accordance with the terms of the Order and any regulation made under it.
- If the cultivator wishes to erect frames, trays or other structure on the sea bed or foreshore he must have the consent of the Crown Estate Commissioners and of the Department of Trade if the structures are hazardous to navigation, and also comply with any planning permission from the local authority.
- In areas designated as Sites of Special Scientific Interest (SSSIs), the cultivator will need the permission of the statutory conservation agency (English Nature) to undertake certain operations. Permission will usually be given unless the operation conflicts with the specific conservation purposes for which the site was designated.

The security of a site needs careful consideration at the onset of site selection since theft by poaching may be a problem in some areas. At present the cultivator has limited legal protection of his stock. Bivalves grown in trays in public waters are protected by the Theft Act, 1968, and the Criminal Damage Act, 1971. Shellfish beds covered by private right of fishery or by Several Order, provided that they are adequately marked, are protected against theft or damage by the provision of section 7 of the Sea Fisheries (Shellfish) Act, 1967.

Several Orders

A cultivator who wants to have additional protection for stock kept in public waters may apply for a right of several fishery. Orders establishing these are granted in the UK under the Sea Fisheries (Shellfish) Act, 1967. They are granted for a fixed period, to an individual, a co-operative or a responsible body, to enable the grantee to cultivate the sea bed within a designated area of water and to conserve, develop and enhance the specified stocks of shellfish thereon. Several rights may also be granted to a Sea Fisheries Committee which cannot cultivate stocks in its own right but may lease rights of several fishery subject to the consent of the Fisheries Departments. The applicant must satisfy the Minister that the fishery will benefit from cultivation. This aspect is monitored by periodic surveys by government scientists, to ensure that the ground is properly cultivated and that the order is being properly put into effect, failing which the right may be terminated.

Regulating Orders

Government may grant a Regulating Order to a responsible body such as a Local Authority, Harbour Board, Sea Fisheries Committee or Fishermen's Association, to enable it to regulate a natural fishery. The fishery may then be fished by the public in accordance with the terms of the order, subject to the observance of any bylaws or regulations made by the controlling body and on payment of any tolls or royalties that may be charged by it.

Private ownership

Private property rights prevent public fishing in some tidal waters. These rights may have been acquired in various ways:

- by grants of rights of a fishery to individuals by the Crown before the Magna Carta (1215);
- by a private Act of Parliament (up to 1868) conferring several rights on boroughs such as Rochester and Colchester or private Companies such as the Seasalter and Ham Oyster Fishery Company Ltd, Whitstable, and the Whitstable Oyster Company;
- by grants of Royal Charter to boroughs;

- by grants of several fishery rights under the Sea Fisheries (Shellfish) Act, 1868, and latterly under the Sea Fisheries (Shellfish) Act, 1967.

Registration of shellfish farms

The Fish Farming and Shellfish Farming Business Order, 1985, has obliged a shellfish farmer to register his business with the Department for Environment, Food and Rural Affairs, the National Assembly for Wales Agriculture Department, or the Scottish Executive Environment and Rural Affairs Department. The purpose of registration is to assist the Departments in dealing with outbreaks of disease if these should occur. Registered businesses are required to keep a record of the stock movements on and off site and to submit a simple summary of movements each year.

Control of deposits

There are certain restrictions on the deposit of shellfish around the coast of Great Britain to prevent the introduction and spread of diseases. January 1993 saw the start of the Single European Market, accompanied in many areas of trade with the removal of internal border controls between Member States. However, movements of shellfish into Great Britain, from both within and outside of the EU, continued to be controlled because of the risk such trade poses to the health of our stocks.

The controls apply to movements to Great Britain of all live molluscan shellfish, their eggs and gametes, from other parts of the EU; from non-EU countries and to deposits within and between coastal zones of Great Britain. The controls operate through a system of:

- approved zones and approved farms, which have a high shellfish health status because they are free from the notifiable shellfish diseases *Bonamia* and *Marteilia*;
- movement documents, which accompany the consignment and are completed by the official fish health protection service where the shellfish are produced;
- pre-notification of movements, which entails written 24 hours' notice to the fisheries department of the arrival of the consignment.

Shellfish from non-EU countries may only be deposited within the EU waters so long as they are certified free from disease by a testing programme as stringent as that which applies in the EU and comply with the other conditions of import. Control of movements of shellfish within Great Britain is based on the occurrence of the disease *Bonamia*. Restrictions on movements currently apply in the areas (Fig. 11.1) between:

- the Lizard lighthouse and Start Point in Cornwall;
- Portland Bill, Dorset, and Selsey Bill in west Sussex;
- Barge Pier, Shoeburyness, in Essex and Landguard Point in Suffolk.

There are no shellfish health controls on moving crustaceans into Great Britain from elsewhere in the EU. There are, however, controls on importing lobsters from third countries and on the deposit of lobsters in the UK, including those obtained from elsewhere in the EU.

Strategic site location

Tactically, the location of a bivalve cultivation enterprise may have important economic consequences for transport and marketing costs. At the local level, site selection should encompass details such as access, since wheeled vehicles and boats are likely to be the workhorses of the enterprise. Over-reliance on manpower is costly and should be kept to a minimum. Not all beach sites are sufficiently firm to allow safe working by heavy vehicles such as tractors. Access to low water or other remote working areas may be possible by boat or light all-terrain vehicles fitted with low pressure tyres. Where a boat provides the only means of reaching a site, then road access to the quay is an important requirement.

Ideally, the shore base should be located adjacent to the sea. This is especially important where there is a need to provide pumped sea water for depuration tanks for cleansing the bivalves free from coliform bacteria prior to marketing. Some modern depuration plants are located some distance from the sea. The need to transport sea water to the site is an extra cost which may not be entirely recoverable by re-using it a number of times or using the alternative but more expensive artificial sea water.

Access to markets is also of major importance to the business enterprise. Live bivalves may be sold to local or national markets, and as a consequence require ready access to an efficient road system to ensure that the product reaches its destination quickly and without loss of quality. Much of the shellfish production in the UK is sold to markets on the continent. This may require ready access to a ferry service and transport via refrigerator lorries where the journey is likely to last for several days. Generally, the relatively low value of market-size bivalves precludes the use of air transport, but the relatively more valuable, small hatchery-reared oyster and clam spat are frequently shipped as air freight to ensure that they reach their destination as quickly as possible.

Bivalves such as mussels and oysters, especially if grown intertidally, are fairly hardy and capable of journey times of several days. Survival is, however, is influenced by a number of factors including the level of rough handling before the journey starts, and temperature and humidity during the journey. Some species of bivalves that normally live subtidally, e.g. scallops, are particularly vulnerable to journey times of more than 60 hours. Specialised misting techniques during transport are being developed to reduce desiccation of the scallops' gills. This is to enable the tissues to receive sufficient oxygen to extend the critical emersion time significantly.

Conflicts of interest within the marine environment

Although bivalve cultivation has taken place for centuries within our estuaries and coastal waters, there is now a growing awareness that other legitimate users may wish to pursue activities that are not always compatible with those of the shellfish farmer. This is especially so with nature conservation interests, usually in estuaries that accommodate significant populations of over-wintering wading birds that feed on the abundant invertebrates to be found on the extensive intertidal mud flats. Other users, such as bait diggers, anglers, boat owners, wildfowlers and commercial fishermen may all wish to use the limited space and resources available to them in our sheltered coast waters. The problem has become sufficiently acute in some estuaries for local authorities to create management strategies that allow the various activities to coexist, in some cases within the bounds of zonation schemes.

In Scotland, the salmon farming industry has been criticised for polluting the environment with some ill-judged placements of sea cages in sheltered, shallow areas with poor water exchange. These conditions have led to an accumulation of debris from fish faeces and uneaten food on the sea bed beneath and around the cage and the consequent smothering of the natural plant and animal communities there. These circumstances may also have a negative impact on the salmon themselves, since anoxic sediments can release toxic gases, such as hydrogen sulphide, and also act as reservoirs of disease organisms. In order to minimise the environmental impact of salmon farms in Scotland, there is now a requirement for the applicant of a new farm to provide an environment impact assessment as one of the conditions of obtaining a lease from the Crown Estate Commission.

This problem is not unique to the fish farming industry. It has also been seen with the shellfish industry, especially with floating mussel ropes or rafts in Spain, Sweden, Japan and Korea, where the effect might be extensive, local or absent depending on the scale of farming and the hydrographical regime of the area. It is essential, therefore, that site selection should incorporate an environmental assessment to forecast the likely impact not only on the marine animals and plants in the area but also on the farmed shellfish as well.

References and bibliography

Anon. (2001) *A Guide to Shellfish Health Controls*. DEFRA, HMSO, London.

Dare, P. J., Davies, G. & Edwards, D. B. (1983) *Predation on Juvenile Pacific Oysters (*Crassostrea gigas *Thunberg) and Mussels (*Mytilus edulis *L.) by Shore Crabs (*Carcinus maenas *(L.))* Fisheries Research Technical Report, MAFF Directorate of Fisheries Research, Lowestoft, **73**.

Hancock, D. A. (1954) The destruction of oyster spat by *Urosalpinx cinerea* (Say) on Essex oyster beds. *Journal du Conseil international pour l'Exploration de la Mer*, **20**, 186–96.

Hancock, D. A. (1955) The feeding behaviour of starfish on Essex oyster beds. *Journal of the Marine Biological Association of the United Kingdom*, **34**, 313–31.

Hancock, D. (1974) *Oyster Pests and their Control*. Laboratory Leaflet, MAFF Directorate of Fisheries Research, Lowestoft (New Series), **19**.

Key, D. & Davidson, P. E. (1981) *A Review of the Development of the Solent Oyster Fishery 1972–80*. Laboratory Leaflet, MAFF Directorate of Fisheries Research, Lowestoft, **52**.

Minchin, D. & Duggan, C. (1989) Biological control of the mussel in shellfish culture. *Aquaculture*, **81**, 97–100.

Spencer, B. E. (1990) *Cultivation of Pacific Oysters*. Laboratory Leaflet, MAFF Directorate of Fisheries Research, Lowestoft, **63**.

Spencer, B. E. & Gough, C. J. (1978) The growth and survival of experimental batches of hatchery-reared spat of *Ostrea edulis* L. and *Crassostrea gigas* Thunberg, using different methods of tray cultivation. *Aquaculture*, **13**, 283–312.

Walne, P. R. (1956) *The Biology and Distribution of the Slipper Limpet* (Crepidula fornicata) *in Essex Rivers*. Fishery Investigations, London, Series 2, **20** (6).

12 Processing live bivalves for consumption

Introduction

The disposal of sewage into the marine environment is not a new or even recent practice in the UK. Since time immemorial it has been an easy option for communities living near to the sea to get rid of their waste by using the regular flushing of the tides to carry wastes away from their doorsteps. Human sewage contains pathogenic bacteria and viruses, which may cause minor or serious illness if passed on to others. Its association with bivalves lies in the latter's ability to filter particulate matter from sea water by the normal filter-feeding process. In this process, bivalves may pump many tens of litres of water per hour through the gills in their body cavity and extract particulate matter, usually microscopic plants and organic detritus, as food.

Although bacteria and viruses themselves are usually too small to be filtered efficiently from the water, they may be bound up with or attached to the surface of other, larger particles in the sea water and are then ingested during the normal bivalve feeding process. Not only are harmful organisms filtered out of the water by bivalves but they are also accumulated and concentrated over a period of time to levels which may be many times greater than those in the water itself. Bacteria and viruses become lodged on the gills and in the gut, from where they may be dislodged quickly by the normal process by which food passes through the digestive tract. Viruses, however, may accumulate within the cells of the digestive gland from where they are difficult to dislodge quickly (Doré & Lees, 1995).

Illness associated with the consumption of bivalves has a long history, but public recognition, and therefore concern, was not aroused until the level of illness reached epidemic proportions. In the UK, the link between gastro-enteric illness, including significantly typhoid fever from which fatalities occurred, and the consumption of oysters and mussels was becoming established in the early part of the 19th century. But it was not until the end of that century that an official investigation was launched (Anon., 1896). The main appendix of this report described the location of cultivated oyster beds and storage ponds in relation to the occurrence of sewer outfalls, and by doing so highlighted the dangers of their close association. The second appendix, by no means less important, dealt with experimental work that described the ability of the typhus bacilli and cholera vibrios to survive in sea water for several weeks and the ability of oysters (native, American and Portuguese oysters) to take up these organisms from the water. Two further appendices provide details of outbreaks of disease in America and France following the consumption of oysters grown in waters contaminated by sewage. Even then, the French recognised that oysters re-laid in clean waters cleansed themselves free of offending organisms.

Perhaps the most infamous disease outbreaks in the UK were those that occurred at two mayoral banquets at Winchester and Southampton in 1902. French oysters, re-laid at Emsworth in storage ponds polluted by nearby sewage outfalls, were served at the banquets, causing gastro-enteric illness in guests who had eaten them raw. Of 267 guests, 119, of whom 117 became ill, ate oysters. Twenty contracted typhoid fever, of whom three died. The seriousness of the situation presented by this example and by the occurrence of numerous other minor incidences of typhoid and gastro-enteric illness before and after the 1902 outbreaks, led the Government to consider ways of dealing with the problem. It was perceived that the options available for solving the problem, were:

- to manage sewage disposal more effectively;
- to close offending fisheries;
- to re-lay the stock into clean waters;
- to sterilise the shellfish by cooking; or
- to purify them from bacterial contamination.

Most of these options would enable the consumer to eat the bivalve raw, the preferred method by many with mussels and oysters. The first two options were discounted on the grounds of cost and politics respectively. The other options were plausible ways of overcoming the problem of bacterial contamination of bivalves. These solutions, however, even today, did not have universal application, since re-laying is expensive and requires reasonably close accessibility of suitable grounds, while cooking of some species of bivalves produces unpalatable changes in the texture of their flesh. The last option eventually was seen as the panacea of the ills of the shellfish industry, but its adoption followed only in the wake of lengthy research to determine the appropriate methodology. It is fortuitous that this research was undertaken on the banks of the River Conwy, with its polluted mussel fishery, since the site eventually evolved into the Fisheries Laboratory, Conwy, which was associated with shellfish cultivation until its closure in 1999.

Mussel purification at Conwy

It is pertinent to mention here that control of the Conwy mussel fishery, which had a history of public health problems in the late 19th and early 20th centuries, was achieved by the imposition of a Regulating Order in 1912. The main requirement of the Regulating Order was that mussels from the fishery should be re-laid in clean waters before they could be sold for public consumption. Since this requirement was not easily attainable, the option of cleansing in local tanks was investigated. Thus, in 1914, four outdoor concrete tanks, pumping facilities and a chlorine plant (Dodgson, 1928) were installed at the site, which for the next 80 years would become known as the Conwy mussel purification tanks.

The site was owned by the Ministry of Agriculture, Fisheries and Food and the purification facility was managed by government staff. The process of purification foreseen at that time was to expose the mussels to sea water containing available chlorine in the form of a dilute solution of sodium hypochlorite, so that the sterilising fluid could percolate within their body cavities. It quickly became obvious that active chlorine could not be used for mussel purification and the method was modified to inactivate the chlorine in the water prior to its use for purification. The procedure required the addition of sodium hypochlorite solution to sea water in a large ($400\,m^3$) storage tank to give a concentration of about 3 ppm active chlorine for *c.* 24 hours. Immediately before use, the sea water was treated with sodium thiosulphate to neutralise the chlorine (8 g sodium thiosulphate per 1000 l of sea water).

The Conwy process of purification

The original four tanks built from reinforced concrete comprised two cleansing tanks for holding mussels, one tank for storing sea water and one tank for chlorinating the sea water. Apart from the division early on, of one of the mussel-holding tanks into two, giving three overall (Fig. 12.1), this system remained unchanged and in operation until 1994. At that time, following the introduction of new European Community Shellfish Hygiene Regulations, which came into force in 1993, the facilities were considered to be too outmoded to meet the new sanitary requirements, and without the large investment of money needed to upgrade them they were closed. The storage and chlorinating tanks, each measuring approximately $22 \times 9 \times 2$ m deep, held $400\,m^3$ of water. The three cleansing tanks were of the same width (13 m) and depth (sloping from 1.1 m at the front to 0.8 m at the back), but were of different lengths, namely 15, 8.5 and 6 m respectively.

Fig. 12.1 Conwy mussel purification tanks.

Their total volume was approximately 370 m^3, sufficient to cleanse 12 t of mussels [200 × 1.25 cwt (62 kg) bags] at each 48-hour cycle of operation. This depuration capacity equates to a low stocking rate of three mussels per 4 l of sea water, equivalent to 32 kg of mussels per m^3 of sea water or 31 kg of mussels per m^2 of floor space.

The floors of the cleansing tanks were constructed with raised, parallel concrete ridges, 10 cm high and 5 cm wide, to support the wooden grids (1.2 × 0.6 m) onto which the mussels were placed at the start of depuration. Fishermen brought their mussels to the tanks after they had been hand-picked or mechanically graded to the legal size limit, and spread them evenly on the grids, 2–3 mussels deep. Tank staff washed the mussels free of mud and loose debris with sea water at high pressure, delivered through a fan jet. The debris passed through the wooden grids into the drainage channels beneath and then to waste. The mussels were covered with *c.* 1.0 m depth of chlorinated sea water with a minimum salinity of 20‰. Surplus chlorine was neutralised by adding sodium thiosulphate to the sea water as it filled the tank. After 24 hours the tank was drained, the mussels were hosed with sterile sea water, and the tank was refilled. After a further 24 hours' immersion, the mussels were again hosed and then covered with a shallow layer of sea water containing active chlorine to sterilise the outside of the shell before packing. The fishermen then shovelled their mussels into 50 or 25 kg sacks, sewed up each of the tops, which were then sealed with a lead seal provided by the Ministry as evidence of purification by the approved manner. Since the mussel fishery is a wintertime occupation, lasting from October to March each year, the depuration tanks were operated at low prevailing temperatures.

Depuration during occasional inclement weather required deviating from the normal operating procedures, but this probably did not happen more than a few per cent of the time. During warmer weather, when sea water temperature in the cleansing tanks exceeded 15°C, the extra requirement for oxygen created by the increased physiological activity of the mussels caused a drop in dissolved oxygen content of the water to below the level of 50% saturation recommended for purification to occur. This led the mussels to close their valves, thereby ending their filtering activity, which in effect ended the purification process. The solution to the problem in recent years was to pump the sea water from the bottom of the tank across its surface to ensure that the dissolved oxygen levels did not fall below the critical level. In cold weather when sea temperatures fell below 5°C, the mussels' physiological processes, and therefore filtering activity, were slowed down sufficiently to reduce the rate of removal of bacteria from their bodies. This problem was overcome by increasing the depuration period, usually by an extra day's immersion in the tanks. These management decisions were aided by the routine bacteriological examination by the local public health laboratory of samples of mussels taken half-way through the depuration process. Experience showed that the mussels reached the required sanitary level of purification set at that time, of 10 *Escherichia coli*/faecal coliforms per ml of flesh, within the first 24 hours of depuration.

The cost of purification was borne by the fishermen, who were charged for each sack of mussels purified (£4 per 50 kg sack, including the cost of the sack, in 1994). The charge, set by government accountants, was adjusted each year as appropriate to offset the cost of providing the service. The cost of purification represented to the fishermen about 17% of the first sale value (£24 per 50 kg in 1994) of the mussels.

The Conwy process of purification did not gain wide adoption, presumably on account of the high capital cost of construction, availability of suitable land and the need for a simpler system capable of dealing with small numbers of shellfish. Tanks based on the Conwy design were built at Lympstone on the River Exe, Devon, in 1924. These were closed owing to the inactivity of the fishery after only a few years of operation. Others were built at Brightlingsea, Essex, for purifying oysters in the 1930s. A simple variant of the method was devised in 1956 for mussel purification in the

River Teign, Devon. This was based on outdoor, shallow (0.6 m depth of water) concrete tanks employing untreated, lightly polluted sea water, in which the mussels were placed for 3 days, with a water change each day (Reynolds, 1956). The search for a small, cheap system began in earnest in the late 1950s in the UK. Germicidal methods other than chlorination were sought, including the use of the powerful oxidising agent ozone and ultra violet light. Government research at the Fisheries Laboratory, Burnham-on-Crouch, Essex, focused on the latter method as being the most likely to produce success (Wood, 1961).

Ultraviolet light purification systems

Early designs

The comprehensive experimental work on the use of UV light as a sterilant of sea water enabled Wood (1961) to prepare a set of advisory instructions for the purification of oysters. The aim of the work centred on producing simple, cheap, purification systems, operable by one person and capable of dealing with small numbers of oysters. He described two systems, one for holding 10 000 oysters (*c.* 1 t) and the other a modular unit with a minimum capacity of 2640 oysters. Both of these systems produced oysters that reached the sanitary requirement within 36 hours of immersion.

The larger unit comprised two adjacent concrete (Fig. 12.2) or wooden tanks, each 6.1 × 1.53 × 0.51 m deep, with a water capacity of 4540 l (i.e. stocking rate of 1 oyster per litre of sea water; 100 g/l). The oysters were placed in a single layer of wooden trays at a density of 540 per m^2, at which they just overlapped. The water was pumped from one end of the tanks into a separate unit fitted with two 30W ultraviolet (UV) tubes and an aerator tower, and back to the other end of the tanks via distribution jets. The pumping rate was half the tank volume per hour. The smaller unit consisted of four tanks with a total volume of 3180 l of sea water, capable of holding 660 oysters per tank. Water was circulated by pumping at half the system volume per hour from a common sump into the UV unit fitted with one 30W tube, into an aerator tower, into the oyster tanks and then back into the sump.

Both the large and small units were stocked with shellfish at *c.* 100 g/l, which was a higher rate than the Conwy chlorination system of *c.* 40 g per litre. This meant that the available dissolved oxygen in the water was consumed at a faster rate in the more densely stocked systems. The design of the UV systems catered for this by incorporating aerator towers, distribu-

Fig. 12.2 Two-tank UV depuration system.

tion jets and a sufficiently fast pumping rate to replenish the oxygen in the water. Pumped water, in addition to aiding aeration, was also necessary to carry the water to the UV sterilising unit, which was housed separately from the oysters. Adequate contact time and low water turbidity were important to ensure adequate sterilisation.

The versatility of these UV systems was further enhanced by incorporating the use of artificial sea water under some circumstances. It was envisaged that artificial sea water salts could be added as a means of topping up salinity, at times when full salinity water was unavailable, to the minimum value necessary for depuration. It also opened up the possibility for the construction of purification tanks inland, away from a ready source of natural sea water. A successful recipe, out of a number of unsuccessful ones tried, consisted of five basic salts (Table 12.1). The same mixture is used today (2001), in slightly different proportions to that suggested by Wood (1961).

The use of these UV purification systems became commonplace in England and Wales. However, there was still a need for a small modular unit, capable of expansion, that occupied limited floor space, was suitable for processing high densities of shellfish in relatively small volumes of water and reduced the cost of using heated water for clams and artificial sea water at inland sites.

Table 12.1 Quantity of salts added to 1000 l of tap water to give artificial sea water at a salinity of 27‰ (Boulter & Denton, 1995)

Sodium chloride	$NaCl$	21.08 kg
Magnesium sulphate	$MgSO_4$	5.18 kg
Magnesium chloride	$MgCl_2$	4.12 kg
Flaked calcium chloride	$CaCl_2$	1.06 kg
Potassium chloride	KCl	0.50 kg

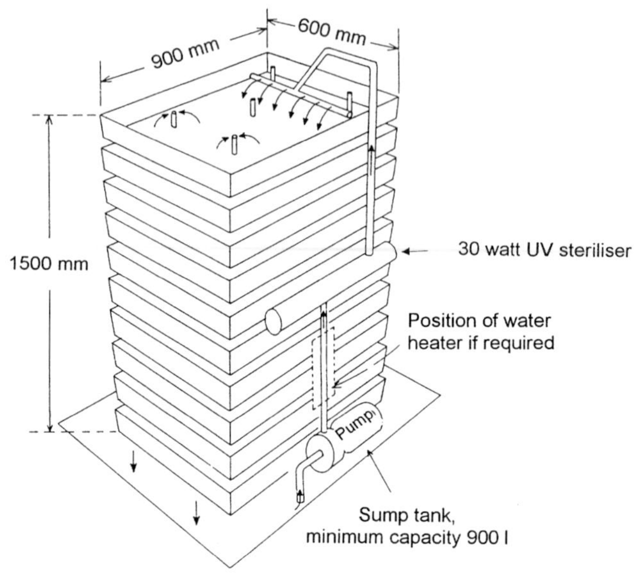

Fig. 12.3 Layout of high density stacking system (adapted from Ayres, 1978).

The development of stacking systems

The solution to the need for economy in floor space and reduced water volume lay in the use of stacking systems. It also meant that the depuration plant could be built indoors with the added advantages of extra hygiene and more precise temperature control than was possible outdoors. The stacking system consisted of a number of individually removable containers that could be slotted into a framework that held them one above the other. In the UK, the prototype design (Fig. 12.3, from Ayres, 1978) showed a system containing ten trays. Since the water reservoir was dead space it could be placed below floor level, from where it was pumped via the UV system and heater (if required) to the top tray. Water then cascaded down from a tray to the one below. The overflow pipes from adjacent trays were staggered to avoid water cascading directly into the overflow pipes below. The trays were also fitted with small drainage holes to allow the water to drain down to the sump reservoir in the event of a power or pump failure.

This is the preferred option, since bivalves are best kept out of water rather than in water devoid of oxygen. Although stocking density for the system is not specifically given (Ayres, 1978), it is assumed that 2500–3000 oysters could be stocked in the ten trays (at 500 per m^2 of tray bottom; three oysters or 300 g per litre of sea water). To ensure that this high stocking density did not compromise the oxygen level in the water (i.e. to keep it above the recommended minimum level of 50% saturation), a pumping rate of one tank volume per hour was recommended.

Operating conditions

The proper management of the depuration plant is essential to ensure that the process achieves the aim of reducing bacteria in the bivalves to a level that makes them safe for human consumption. Plant design, therefore, should specify stocking capacities, flow rates and other management requirements to ensure that oxygen levels exceed the minimum necessary to support the normal pumping and filtration activity of the bivalves. Since the management requirements differ between species of bivalves it is necessary for the depuration strategy to be altered accordingly.

Salinity and temperature

There are seven species of bivalves that are routinely (oysters and mussels), or less commonly (clams, razer clams and cockles), purified in the UK. The minimum salinity and temperature ranges recommended for use in depuration to attain normal functioning of these species are shown in Table 12.2.

In the UK, the salinity of natural sea water rarely exceeds 35‰ but frequently falls below this value depending on location and rainfall. Artificial sea water salts, using the recipe given earlier, may be used to increase salinity at times when natural sea water does not meet the minimum values required. Artificial sea water salts are , however, relatively expensive and may be suitable only for use in small depuration plants.

Sea temperature fluctuates widely seasonally and diurnally. Temperatures outside of those shown in Table 12.2 may prove troublesome for depuration plant management. Bivalves respond to low temperatures by slowing down their physiological process to an extent that the filtration level is insufficient

Table 12.2 Recommended minimum salinity and temperature ranges for different species

Common name	Latin name	Salinity (‰)	Temperature (°C)	
			Minimum	Maximum
Flat oyster	*Ostrea edulis*	25	5	15
Pacific oyster	*Crassostrea gigas*	20	8	18
Mussel	*Mytilus edulis*	19	5	15
Palourde	*Tapes decussatus*	25	12	20
Manila clam	*Tapes philippinarum*	25	12	20
Hard-shell clam	*Mercenaria mercenaria*	20	12	20
Cockle	*Cerastoderma edule*	20	7	20

for adequate expulsion of bacteria from their bodies. High temperatures cause excessive bivalve physiological activity and oxygen consumption, resulting in reduced oxygen levels in the water. These conditions cause the bivalves to isolate themselves from the inclement conditions by closing their shells, thus effectively bringing the depuration process to an end. At some times of the year, therefore, temperature control may be necessary to ensure that the water temperature is kept within the range required for successful depuration. Modern depuration plants are usually fitted with heaters but less commonly with chillers that are relatively costly to buy and to run.

UV light

Ultraviolet light is the most commonly used method of sterilising sea water in depuration plants in the UK, USA and Australia. The bactericidal properties of UV light lie in the emission of radiation within the wavelength range of 250–260 nm, which irreversibly damages the DNA of living organisms. Close contact between UV light and the water is essential to obtain the necessary sterilisation. The employment of recirculated water in depuration plants makes it unnecessary to use the high pressure mercury lamps used for single-pass water sterilisation in the preparation of potable water. Most systems in the UK use 30W low-pressure mercury vapour lamps which emit much of their UV energy within the sterilising wavelength range. The installed capacity of these units is 8.2W of energy per 1000 l of water with the required wattage supplied as single or multiple 30W lamps according to the system volume. In older depuration plants, the water was pumped across a weir to ensure close proximity with the UV lamps that were housed in a unit separate from the bivalves. In modern plants the UV lamps are immersed in the water within a quartz glass tube fitted inside a plastic column through which the water flows. Since the depuration process depends on adequate penetration of the UV light, relatively clean water with a low particle content must be used. Also, it is necessary to keep the lamps clean, and to change them after a specified period (usually 1000 hours), since their efficiency, and therefore sterilising capability, decreases with use.

Modern UK depuration plants

The stacking system developed by Ayres in the 1970s, because of its compactness, gained much popularity with the UK industry, which dealt with relatively small quantities of bivalves compared with its continental neighbours. Important changes in design took place between the mid-1980s and 1996 when the Sea Fish Industry Authority (SFIA) assumed responsibility for the development of depuration plant technology. Their philosophy was to introduce flexibility into plant design to enable it to cope with quantities of bivalves ranging from a few hundred kilograms up to 10 t. The plants were also engineered in such a way to ensure that their operation produced an end product of the required sanitary standard without the need for regular bacterial testing of the bivalves.

The SFIA developed two types of modular depuration plant design within which the base unit could be scaled up to incorporate multiple systems. They retained the idea of the small stacking system suitable for holding *c.* 200 kg of bivalves and modernised it to incorporate purpose-made individual trays. This was called the vertical stack system. Larger plants, with the basic units capable of holding 750 and 1500 kg of bivalves, were also developed. The latter systems retained compactness by keeping the bivalves at high densities in containers that largely filled the width and depth of the tank. These were called multi-layer tank systems.

The vertical stack system

This system is similar to that devised by Ayres (1978). The base unit (Anon., 1995a) comprises a sump tank with a volume of 650 l, capable of holding the drained water from the 16 trays above. The specially designed plastic trays (80 × 45 × 15 cm), arranged in two columns on a stainless steel frame, are free-standing and easily removable to gain access to the stock. Water pumped from the sump to the top trays cascades down through staggered overflows. They are designed to cause minimum disturbance to the stock and to create laminar flow across the tray between the inlets and outlets. The cascading effect of water tumbling between trays together with a flow rate of 900 l per hour (*c.* 1½ unit volume per hour) is necessary to main-

Table 12.3 Recommended stocking capacity, density and flow rates for the stack system (Anon., 1995a)

Species	Stock per tray (kg)	Stock depth	Density (g/l)	Specific flow rate (l/kg/hour)
Mussels	15	80 mm	370	3.8
Flat oysters	12.5	one deep	308	4.5
Pacific oysters	12.5	two deep	308	4.5
Clams	21	80 mm	517	2.7
Cockles	15	80 mm	370	3.8

tain oxygen levels above the minimum of 50% saturation required for the recommended stocking capacity of the unit. The recommended stocking capacity of the unit, depending on species, is shown in Table 12.3.

The 25W UV unit and vulnerable electrical components, are located behind the stacks of trays, protected from splashing by a plastic shield. The unit is designed to operate within the salinity and temperature ranges shown (Table 12.2) and may be fitted with a thermostatically controlled heater and chiller to achieve this.

The compactness of the unit made it suitable for use with artificial sea water (Table 12.1). Trials by SFIA showed that the water could be re-used for up to 30 days with 10% replacement after each depuration cycle. This topping up was necessary to replace that lost by spillage, evaporation and discard of residual water containing settled faeces and pseudofaeces from the bivalves, left in the trays after drainage into the sump.

Although the base unit held 200–300 kg of bivalves, the simple modular design enabled it to be expanded into larger units similar to that containing 200 trays, with a capacity of 2 t of oysters (Fig. 12.4).

The multi-layer systems

The base units of the medium and large scale multi-layer systems, each holding 2600 and 9200 l of sea water respectively, are designed for purifying 750 kg and 1500 kg of mussels (Anon., 1995b,c). Each unit consists of a single, marine grade stainless steel or glass reinforced plastic tank (medium system only) (Fig. 12.5).

The bivalves are placed in plastic mesh containers that allow relatively free circulation of water through the stock. With containers measuring 60 × 50 × 18 cm deep, 48 can be stacked six high in the medium, and 92 in the large multi-layer tank. The recommended stocking capacity of a container of these dimensions is the same as that shown for the trays in the vertical stacking system (Table 12.3). The containers are placed between two verti-

Fig. 12.4 Large stacking system for oyster depuration.

Fig. 12.5 750 kg multi-layer mussel purification tank. (Photo: D. B. Edwards).

Table 12.4 Stocking density and flow rates of the medium and large multi-layer systems (Anon., 1995a)

System	Capacity (kg)	Volume (litres)	Stock density (g/l)	Flow (m^3/hour)	Specific flow rate (l/kg/hour)
Medium	750	2600	288	12.5	16.7
Large	1500	9200	163	9.2	6.3

cal, perforated flow screens, one at each end of the tank, which produce a laminar water flow to ensure that oxygenated water reaches each level of containers. The stocking densities and flow rates used in these systems are shown in Table 12.4.

The high flow rate of the medium size system (five unit volumes per hour) provides sufficient oxygen to permit its use within the full range of recommended temperatures for mussels (5–15°C; Table 12.2). The relatively low flow of the large size system (one unit volume per hour), however, permits it's use only at 5–10°C. Operation at 10–15°C is achieved by injecting filtered air into the sea water. UV light sterilisation is used for these systems, with 1 × 30W and 6 × 30W lamps incorporated into the medium and large scale systems respectively.

The modular design of these units enables them to be linked together. Scaled up versions with suitably modified UV and pumping facilities for purifying up to 10 tonnes of bivalves have been constructed.

All the modern UV units developed by SFIA are designed to produce cleansed bivalves after 42 hours of operation. The proper management of the plant is essential to consistently achieve the aim of removing harmful bacteria from the bivalves. The information contained in this chapter provides a bare outline of procedures. More detailed information on handling of bivalves before depuration, re-use of sea water and artificial sea water, excessive foaming of sea water and other problems associated with the process, is provided in the SFIA operating manuals.

Approval of purification systems

The handling of live molluscan bivalves from harvesting to retail sale is governed by the Food and Safety (Fishery Products and Live Shellfish) (Hygiene) Regulations, 1998. Before purification plants are able to operate, they must be issued with an approval number, given by the Food Authority, only if the requirements of the Regulations are met. In addition,

Table 12.5 Number of licensed depuration plants in UK (1998)

England & Wales	48
Scotland	20
N Ireland	2
Total	70

a 'conditions of approval' document is issued for each purification system, specifying the conditions of usage.

In 1998, there were 70 licensed depuration plants in the UK (Table 12.5). All were UV systems, mostly based on the modern Sea Fish Industry Authority design, but with a few of the older, double concrete tank installations still in use. Of all installations, about 10% used artificial sea water.

In addition to approving plants for depuration, the regulations also ensured that each bag of purified shellfish was tagged with information to enable it to be traced back to the plant in the event of a disease outbreak. Additional records at the plant provided the authorities with further information, such as the date, fisherman and harvesting area of the shellfish, to allow consideration by whether fishing activities should be restricted.

Classification of bivalve harvesting areas in the UK

The European Council Directive (EEC 91/492), concerning the health conditions for the production and placing on the market of live bivalve molluscs, are enforced in the UK under the Food and Safety (Fishery Products and Live Shellfish) (Hygiene) Regulations, 1998. The Directive required member countries to classify their shellfish production areas according to the extent to which shellfish sampled from them were contaminated with *E. coli* or with faecal coliforms. The standards set for the four categories are shown in Table 1.5, Chapter 1. Shellfish from grade A areas were considered to be sufficiently clean to be sold for human consumption without further treatment, whereas those from grade B and C areas required further treatment. No harvesting was permitted from prohibited areas. Approved treatments included depuration, cooking and re-laying, and were expected to reduce the level of contamination to that of grade A shellfish. Other end-product standards were set to ensure that *Salmonella* and algal toxins caused no problems to public health.

Collection of data on which shellfish harvesting waters in the UK are classified is the responsibility of the Food Authority. Comprehensive seasonal sampling of bivalves from the beds provided information on the coliform

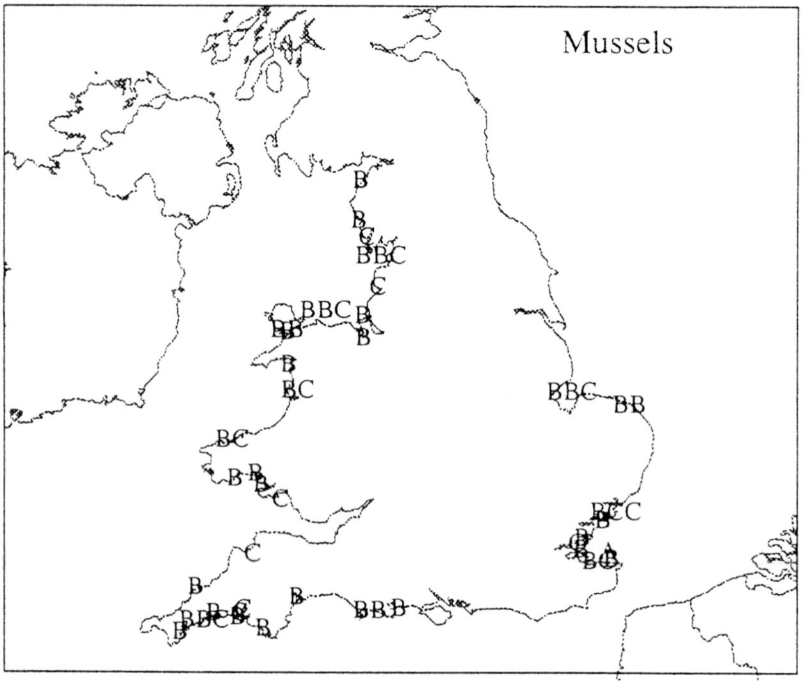

Fig. 12.6 Classification of mussel beds in England and Wales in 1999.

(*E. coli*) levels in the flesh. These data, evaluated by staff at the CEFAS (Centre for Environment, Fisheries and Aquaculture Science (formerly MAFF), Fish Disease Laboratory, Weymouth, were used to produce a classification for England and Wales. In 1999, 207 classifications were made, of which 5% were grade A, 69% grade B, 23% grade C and 3% grade D. The distribution of classified mussel beds in 1999 is shown in Fig. 12.6. The classification is reviewed each year, and depending on additional information gained on the status of a bed its classification may be changed either upwards or downwards. Fortunately the introduction of the Shellfish Waters Directive now requires the water companies, who manage the discharge of sewage into the sea, to improve the quality of shellfish production waters.

Other methods of cleansing shellfish

Although UV light is the preferred method of sterilising sea water in depuration plants in the UK, other sterilising methods, similar in concept, are used in France, Spain and Italy. These all entail placing harvested bivalves in shore-based tanks containing circulating water sterilised by ozone, iodoform or gaseous chlorine. Natural cleansing of bivalves by re-laying stock onto beds with cleaner waters is also an option for the cultivator. Processing shellfish such as cockles by heat treatment is a method used traditionally to render them fit for consumption.

Re-laying

Re-laying shellfish into cleaner waters must be done for a sufficiently long period of time to ensure that they purge themselves free of the heavier load of bacteria acquired on the previous beds. The time required to achieve this aim is dependent on the level of contamination, the season and conditions prevailing during the period of re-laying. The current regulations (see above) require re-laying beds to be approved by the Food Authority by the testing process applied to the classification of shellfish harvesting beds. Lightly contaminated shellfish may require relatively short re-laying periods in ideal conditions, but those from beds classified as grade C, must be re-laid for 2 months (Table 1.5, Chapter 1).

From the practical point of view, re-laying is expensive since it involves extra handling of the shellfish with, perhaps, some mortality involved. Also suitable, available beds, nearby, with a lower classification, may not be readily available to the shellfish grower.

Heat treatment

Live cockles are rarely sold to the UK market nowadays, partly because transport costs are relatively high for this low value species. Processing cockles by cooking is a traditional method of preparing this species for market, since transporting cockle meats to market is a more viable option. The cooking process must effectively kill off bacteria contained in the flesh, yet produce meat that is tender, succulent and highly marketable. Under-cooking produces meats that are soft and jelly-like, which do not stand up to salting or handling without tearing. Excessive cooking of cockles and mussels, however, can ruin their texture and reduce their volume by up to 20% and thereby adversely affect their marketability. Traditionally cockles are cooked by boiling or steam pressure-cooking.

The process came under close scrutiny in the 1980s following disease outbreaks arising from the consumption of cockles from the Thames Estuary fishery. The former MAFF Food Science Laboratory at Torry, Aberdeen, in conjunction with the Virus Research Laboratory (Public Health Laboratory

Service), undertook extensive trials to determine a standardised process that achieved the aims of sterilising the meats without unduly altering the taste of the flesh (Early & Nicholson, 1994). The process that was developed was based on the inactivation of the Hepatitis A virus. It was essentially a continuous process that allowed for a controlled volume and depth of shellfish to pass through water, at a controlled temperature, for a controlled period of time. The aim was for the shellfish meats to be held for at least 90 seconds at 90°C, conditions which were sufficiently stringent to kill many types of virus. Various safeguards and recording devices were incorporated into the system to ensure that these times and temperatures were maintained. The development of this process preceded, yet fully met and exceeded, the end-product standards, based on bacterial levels of the flesh, of the EC Directive of 1991. All modern heat treatment plants in the UK are based on the Torry method and must be approved by the Department of Health before they are licensed to operate.

Ozone depuration

Following earlier studies that showed its disinfectant properties in sea water, ozone was first used in depuration plants in the early 1960s in France. It was so successful that it has become the depuration method employed in all the major cleansing stations in France. Ozone, the triatomic form (O_3) of oxygen, is a powerful oxidising agent, and kills bacteria and viruses in the water with which it comes into contact. Ozone is produced by a generator, which is expensive to buy and to run. As with UV light, ozone is introduced in a contacting chamber separate from the shellfish, at a concentration of 0.5–1.5 mg/l and with a contact time of 10–20 minutes. Residual ozone is removed, using compressed air or by allowing the water to cascade into in a degassing chamber, which results in a beneficial enrichment in oxygen content of the sea water. Excessive amounts of ozone can lead to the production of significant amounts of hypobromous acid in the water, which interferes with the normal filtration process of bivalves and causes them to close their shells.

Iodoform depuration

Iodoform disinfection of sea water in depuration plants is not widely practised but has gained some popularity in Italy. The sterilising agent is a substance consisting of a complex of iodine and hydriodic acid (*Actomar*® by Ciba-Geigy) which produces a strong disinfecting action. The iodoform is dosed into the

Table 12.6 Advantages and disadvantages of the different methods of depuration (from Casagrande, 1988; Morel & Salamon, 1992; Rodrick & Schneider, 1992)

	UV light	Ozone	Chlorine	Iodoform
Capital costs	low	high	medium	medium
Operating costs	low	high	low	low
Installation	simple	complex	complex	complex
Maintenance	low	high	low	low
By-products	none	toxic	THM*	THM*
Dose	25 mW/s/cm^2	0.5–1.5 mg/l	3 mg/l	0.5–2 mg/l
Contact time	1–5 seconds	10–20 minutes	30–60 minutes	30–60 min.
Toxic to shellfish	no	yes	yes	no

THM* = trihalomethanes

sea water at 0.5–2 mg/l without the need to separate the chemical from the bivalves. It is claimed that the direct contact between shellfish and sterilant reduces depuration time to *c.* 8 hours (Casagrande, 1988).

Depuration using chlorine gas

The use of chlorine gas as a sterilant for bivalve depuration is popular in Spain and Italy and to a much less extent in France. Because of the small quantities required to achieve sterilisation, equipment capable of injecting the gas precisely into the water is needed. The gas is injected in a compartment separate from the bivalves. Surplus chlorine is removed by passing the water through a bed of activated charcoal. A comparison of some aspects of each of the main depuration methods is shown in Table 12.6.

References and Bibliography

Anon. (1896) *On the cultivation and storage of oysters and certain other molluscs in relation to the occurrence of disease in man – Report to Government on the status of oyster fisheries – Annex I* by H. T. Bulstrode: in: Report and Papers Submitted by the Medical Officer of the Local Government Board.

Anon. (1995a) *Seafish Standard Design Purification Systems: Operating Manual for the Vertical Stack System*. Technical Information Service, Sea Fish Industry Authority, Hull, England, Report, Reference No. 1995/32/FT-V1.

Anon. (1995b) *Seafish Standard Design Purification Systems: Operating Manual for the Medium Scale System*. Technical Information Service, Sea Fish Industry Authority, Hull, England, Report, Reference No. 1995/31/FT-V1.

Anon. (1995c) *Seafish Standard Design Purification Systems: Operating Manual for the Large Scale Multi-layer System*. Technical Information Service, Sea Fish Industry Authority, Hull, England, Report , Reference No. 1995/33/FT-V1.

Ayres, P. A. (1978) *Shellfish Purification in Installations Using Ultraviolet Light*. Laboratory Leaflet, MAFF Directorate of Fisheries Research, Lowestoft, **43**.

Ayres, P. A. & Wood, P. C. (1977) *Artificial Sea water for Shellfish Tanks, Including Notes on Salinity and Salinity Measurement*. Laboratory Leaflet, MAFF Directorate of Fisheries Research, Lowestoft, **39**.

Boulter, M. & Denton, J. W. (1995) *Development of Small-scale Vertical Stack Oyster Purification Plant with Repeated Re-use of Artificial Sea Water.* Seafish Report, **459**, Sea Fish Industry Authority, Hull, England.

Casagrande, F. (1988) Effective purification in aquaculture. *Marine Farming*.

Dodgson, R. W. (1928) *Report on mussel purification*. Fishery Investigations, Series II, **10** (1).

Doré, W. J. & Lees, D. N. (1995) Behaviour of *Escherichia coli* and male-specific bacteriophage in environmentally contaminated bivalve molluscs before and after depuration. *Applied and Environmental Microbiology*, 2830–4.

Early, J. C. & Nicholson, F. J. (1994) *Specification for a Model Cockle Processing Plant.* Central Science Laboratory, Food Science Laboratory, Torry, Unpublished Report, TD 2401.

Morel, M. & Salamon, W. (1992) *Project of a Depuration Plant in Northern France.* Purification des Coquillages. Deuxième Conference International, April 1992, IFREMER, Rennes, France, 323–36.

Reynolds, N. (1956) *A simplified system of mussel purification*. Fishery Investigations, Series II, **20** (8).

Rodrick, G. E. & Schneider, K. R. (1992) Depuration and re-laying of molluscan shellfish. In: *Environmental Indicators and Shellfish Safety*. (eds C. R. Hackney & M. D. Pierson), pp. 331–63. Chapman and Hall, New York.

Wood, P. C. (1961) *The principles of water sterilization by ultra-violet light, and their application in the purification of oysters*. Fishery Investigations, Series II, **23** (6).

Glossary

Adductor muscle(s)	Muscle(s) responsible for closing the shells of bivalves.
ASP	Amnesic shellfish poisoning induced in humans who eat bivalves and containing toxins from naturally occurring single-celled marine plants.
AZP	Azaspiracid shellfish poisoning, induced in humans who eat bivalves containing toxins from naturally occurring single-celled marine plants.
Biodeposits	Waste material arising from bivalve faeces and pseudofaces that accumulate on the beds by the filtering activities of mussels and oysters etc.
Bivalve	Mollusc having two shell valves.
Bouchot	Vertical pole driven into intertidal ground for growing mussels in France.
Byssus	Protein thread secreted by a gland, for attaching some species of bivalve to substrate (mussel, larval clams and scallops).
CEFAS	Centre for Environment, Fisheries and Aquaculture Science. An Executive Agency for DEFRA.
Cephalopod	Mollusc such as squid, cuttlefish and octopus.
Claire	Shallow pond used in France for fattening oysters and clams.
Coliform bacteria	Bacteria found in the gut of humans and some mammals.
Crystalline style	Component of stomach of herbivorous molluscs, which aids digestion by mechanical and enzymatic activity.
Cultch	Material, usually shells, used for the collection of spat.
Cultchless spat	Hatchery-reared spat removed from cultch soon after settlement or induced to settle in the absence of cultch.
DEFRA	Department for Environment, Food and Rural Affairs.
Depuration	Process of purging harmful bacteria from live bivalves, rendering them safe for human consumption.
Diatomite	Fine powder made from diatoms used for filtering particles from sea water in the hatchery process.
Dioecious	Having separate sexes.
D-larva	Characteristically D-shaped, early larva of scallop, clam and Pacific oyster.

DSP	Diarrhetic shellfish poisoning induced in humans who eat bivalves containing toxins from naturally occurring single-celled marine plants.
Ectoparasite	Parasite living on the outside of the body of the host.
Flupsy	Floating upwelling system.
Foot	Muscular organ of mollusc enabling it to crawl (whelk, abalone), burrow (clams) or plant the byssus (mussel). May be vestigial or absent in some bivalve species (scallop, oyster).
Gamete	Mature sex cell within the testis and ovary or newly released, prior to fertilisation.
Gastropod	Mollusc with univalve or reduced or absent shell (whelk, abalone, slug).
Gill (ctenidium)	Ciliated organ that creates a water current through the mantle cavity of molluscs for ventilation and/or collection of microscopic food by filtration.
Glycogen	Carbohydrate reserve of body.
Gonad	Male or female sex organ.
Haemocytometer	Glass counting chamber used for estimating phytoplankton numbers.
Hermaphrodite	Having both sexes simultaneously.
Heteromyarian bivalve	Bivalve in which the two valves are adducted by a small anterior and a larger posterior muscle, e.g. mussel.
Hydroid	Sessile animal, in some colonial species with plant-like form having a hard chitinous exoskeleton favoured by mussels as a settlement surface.
Imposex	Abnormal development of male organs in female marine snails, caused by TBT leachates in the water.
Isomyarian bivalve	Bivalve in which the two valves are adducted by two muscles of similar size, e.g. clam.
Larvipary	Name of condition in bivalves that incubate their eggs in the mantle cavity, eventually spawning them as larvae.
Lipid	Fat reserve of body.
Mabe	Name given to a semi-spherical pearl derived by inserting a shaped nucleus between shell and mantle of abalone.
Macrophytic algae	Multicellular algae having large fronds.
MAFF	Ministry of Agriculture, Fisheries and Food (England and Wales). Renamed the Department for Environment, Food and Rural Affairs (DEFRA) in 2001.
Mantle	Mollusc tissue enclosing the organs of the body and which secretes the shell.
Mesocosm	Semi-artificial environment used for studying behaviour of animals and plants under semi-controlled conditions.
Metamorphosis	The process of changing from larva to juvenile form.

Microalgae	Usually refers to single-celled algae or those with a few cells in chains.
Monomyarian bivalve	Bivalve in which the two valves are adducted by a single, central muscle, e.g. scallop, oyster.
Nacre	Layer of mother-of-pearl on internal surface of mollusc shell.
Operculum	Chitinous apparatus for sealing shell opening in some species of gastropods.
Pediveliger	Late larval stage in which larva uses its foot to crawl along the bottom seeking a surface suitable for settlement and metamorphosis.
Periostracum	Horny outer layer of shell of bivalves.
Phagocytosis	Engulfment of food or waste particles by specialised cells within the digestive or vascular system during digestion or excretion.
Phytoplankton	Plant plankton.
Plankton	Small plant or animals that drift with the currents in the sea.
Pôche	Plastic mesh bag for growing bivalves.
Poll	Part of a fjord in Norway used for breeding and growing oysters.
Polychaete worms	'Bristle' worms with many appendages.
Polyculture	Cultivation of a number of complementary species simultaneously in the same system.
Protandry	Maturing first as a male.
Pseudofaeces	Uneaten waste particles aggregated on the gills and mantle during filter feeding and discharged via the inhalant aperture.
PSP	Paralytic shellfish poisoning induced in humans who eat bivalves and some crustaceans containing toxins from naturally occurring single-celled marine plants.
Psu	Practical salinity unit [equivalent to parts per thousand (‰)].
Pufa(s)	Polyunsaturated fatty acid(s).
PVC	Polyvinyl chloride.
Radula	Mouthpart of some herbivorous and carnivorous gastropods used for rasping algae from rocks or drilling holes in bivalve shells to gain access to the flesh.
Ren	Synthetic or wire string threaded with oyster or scallop shells used for collecting and growing spat in Japan.
Retractor muscles	Muscles responsible for retracting the foot into the mantle cavity.
Seed	Spat sufficiently large for re-laying in the sea.
SFIA	Sea Fish Industry Authority.
Spat	Post-larval bivalves up to *c.* 12 months.
Spatfall	Newly-settled young bivalves following metamorphosis of the larva.
TBT	Tributyl tin, an additive to paint as a marine boat antifoulant.

Trestle	Intertidal frame of wood or metal to support oyster/clam trays off the ground.
Triploid	Artificially induced condition in bivalves to produce three sets of chromosomes rather than the normal two.
Trochophore	Early swimming stage of mollusc larva before shell development.
Univalve	Molluscs having one valve (whelk, abalone).
UV	Ultraviolet.
Velar retractor muscles	Muscles responsible for retracting the velum into the mantle cavity of a larva.
Veliger	Swimming larva.

Index